Technology Journal

With
Lesson
Plans

SRA

REAL SCIENCE

D1370181

William C. Kyle, Jr.   **Joseph H. Rubinstein**   **Carolyn J. Vega**

SRA

A Division of The McGraw-Hill Companies

Columbus, Ohio

Authors

William C. Kyle, Jr.
E. Desmond Lee Family
   Professor of Science Education
University of Missouri – St. Louis
St. Louis, Missouri

Joseph H. Rubinstein
Professor of Education
Coker College
Hartsville, South Carolina

Carolyn J. Vega
Classroom Teacher
Nye Elementary
San Diego Unified School District
San Diego, California

PHOTO CREDITS
Cover Photo:
Bengal Tiger
© Tim Davis/Tony Stone Images

SRA/McGraw-Hill

A Division of The **McGraw·Hill** *Companies*

Send all inquiries to:
SRA/McGraw-Hill
250 Old Wilson Bridge Road
Suite 310
Worthington, Ohio 43085

Printed in the United States of America.

ISBN 0-02-661345-X

1 2 3 4 5 6 7 8 9 VIC 05 04 03 02 01 00 99

Table of Contents

Scientific Literacy for Today's Media-Savvy Students

Welcome to the *SRA Real Science* **Technology Journal**—a unique guide, specially designed to help you bring science to life through video, hands-on activities, cross-curricular connections and various informal assessment options. Featuring ready-to-teach video lesson plans and student blackline masters, this book provides everything you need to use technology effectively.

Through your consistent use of the **Technology Journal**, your students will begin to build a solid foundation of knowledge and experiences about life, earth, physical and health science. At the same time, they will begin to master the scientific processes necessary to solve problems in today's electronic world.

A Picture's Worth a Thousand Words

More than 90 percent of today's students are visual learners. Television, computers and video games have helped to define this generation. *SRA Real Science* is designed to appeal to today's media-savvy students. Three years of longitudinal studies have shown that in science classrooms where video and technology are used in a specific, consistent manner, there is:

- Nearly universal increase in student motivation and achievement among students with varying backgrounds and ability levels
- Magnified effect in students who had shown less initiative with more conventional academic approaches
- Ability to handle more complex assignments with higher-order thinking skills

The program builds visual comprehension by developing more discerning viewers.

Multiple Approaches

But seeing isn't everything. Research has shown over and over again that mentioning or merely showing something does not ensure student understanding. Although a few students may be interested enough to remember the concept at hand, more often than not, information provided in isolation is lost. For students to understand and retain the information, teachers must present concepts in context and through a variety of experiences. Teaching for concept mastery requires multiple exposures to the same concepts. The *SRA Real Science* **Technology Journal** is structured to help you do just that. As a result, your students will get more out of what they see, do and hear.

ClearPath Lesson Plans

Today's classroom teacher faces a lot of demands. The *SRA Real Science* **Technology Journal** is a practical choice that works well in classrooms where time constraints are a reality. The Program features SRA's unique ClearPath Lesson Plans. These lesson plans offer a systematic, straightforward approach to presenting lesson content and are specifically developed for teachers with varying degrees of science background. Each ready-to-teach lesson plan is organized in a simple, three-step format to provide you with the support that you need in the time you have to teach:

❶ Introduce

Lesson Background is a science content resource for you. With all you're expected to know, you can't know everything about science. This paragraph gives you a quick overview of the content covered in the lesson. **Activating Prior Knowledge** will help you understand what your students think about a concept—with clear explanations of how to interpret student understanding and to address misconceptions.

❷ Teach

This section provides detailed information for using each video and accompanying **Video Notes** and **Activity** blackline master pages. Also included is guidance for setting learning goals before students watch the video, a quick check for comprehension after they view the video, and a structured note-taking strategy using **Video Notes** worksheets.

❸ Close

The last section provides assessment opportunities through a detailed **Checkpoint**, designed to verify that the learning objectives have been met.

Other features of the ClearPath Lesson Plan:

The Big Question

Each Video Lesson presents a Big Question that emphasizes key science concepts that support the AAAS Benchmarks, the National Science Education Standards and state science guidelines.

Objectives and Vocabulary

Lesson objectives and featured vocabulary words presented through The Big Question are located in the left column of each lesson plan.

Using Different Media

The video portion of the program is available in videocassette, laserdisc and CD-ROM formats. Information for accessing the video for each delivery-system is located in the left column of each lesson plan.

Fun Facts

Teachers can become "instant experts" about the featured content through these interesting facts located in the right margin of each lesson plan.

Hands-on Activities

Exploratory and confirming activities help students practice science process skills. "What to Expect" helps the teacher manage the activity. Blackline masters for the activities immediately follow the lesson plan pages. Select lessons feature Alternate Activities for additional hands-on opportunities.

Across the Curriculum

Integrate science with other content areas through this feature, located in the right margin of each lesson plan.

Safety in Science

Safety in the classroom is one of a teacher's primary concerns. The activities included in this program avoid the use of hazardous chemicals and dangerous procedures. Establish behavior that you expect from your students at all times while doing science activities. Most importantly, model correct safety procedures at all times for students.

The Next Generation in Elementary Science

The *SRA Real Science* **Technology Journal** is carefully crafted for today's classroom. Students get a media-enhanced presentation of essential science knowledge, vocabulary and skills, including life, earth, physical and health science at each grade level. Teachers enjoy a straightforward program that does not expect them to be science experts—and helps them better appeal to today's media-savvy students. It's a scientific success for everyone.

Organisms & Where They Live
Habitats of Organisms

The Big QUESTION *How do animals and plants live where they do?*

Media Resources
Videotape
Grade 3 • Life Science

Laserdisc
Grade 3 • Life • Side 1
SAMPLE
Frame 12345

1 Introduce

Lesson Background An environment is made of all of the living and nonliving things that surround an organism. A plant or an animal's address is its habitat, which provides everything that it needs to live (food, shelter, air, water). A group of one type of animal or plant is called a population, whereas a group of different kinds of animals and plants sharing one area is called a community. The term ecosystem applies to all living and nonliving things and the relationships between them. Students will easily relate to a population of squirrels living in their habitat in the forest, which provides them with plenty of food and shelter. A squirrel's community includes oak trees, worms and birds, side by side with many other plants and animals. These living things, along with nonliving things such as rocks, make up the ecosystem of the forest.

Activate Prior Knowledge Ask students to define the term environment and provide examples of different environments. Write their responses on the chalkboard.

2 Teach

Before Viewing the Video
For Discussion

Ask students to name plants and animals that they might find in the environments they just named. Ask students to describe ways that organisms' environments help different organisms survive, e.g., What do they eat? How do they get their food? How do they hide from predators? Record their responses on the chalkboard. Show the video and tell students to pay close attention to information about organisms and their habitats.

Objectives
- distinguish between environments and habitats
- describe how organisms meet their needs
- identify the composition of ecosystems
- define community and population

After Viewing the Video
Thinking Critically
1. Ask students to recall habitats and organisms that they saw in the video. Add these examples to the lists on the chalkboard.

2. Have students name nonliving things that are found in ecosystems. How are these things beneficial to living things?

Using Video Notes

Distribute copies of Page 8 to students. Give students several minutes to review the terms and statements on their worksheet. Show the video again. Have students match the terms to the statements and use these terms to describe the surroundings of a specific organism.

Vocabulary
organisms, environment, habitat, ecosystem, community, population

❸ Close

Checkpoint

- What is an environment? What is a habitat? (An environment is made of all the living and nonliving things that surround an organism. A habitat is where an organism lives and grows.)
- How do organisms meet their needs? (Organisms are adapted to the places they live.)
- What are ecosystems composed of? (Ecosystems include all living and nonliving things and their habitats.)
- What is a community? What is a population? (A community is made of living things in an area. A group of the same kinds of living things in an area is a population.)

Activity

What to Expect Students will compare the habitats of different communities by creating their own ecosystems.

Distribute copies of Page 9. Encourage students to add other plants and animals that would live in each community. Also, make sure that their habitats include food and shelter.

Alternate Activity

Find Out Why do different plants and animals live in different habitats and communities?

What You Need 10-cm x 15-cm note cards, markers

What to Do

1. Provide each student with the name of an organism and have him or her research the organism's habitat and adaptations, writing these details on one side of a note card.
2. Set up students in a game show where a panel of three "experts" is charged with guessing the name of each contestant's identity.
3. The panel of experts will ask yes-or-no questions about the organism's habitat and adaptations and the contestants will give information from their note cards. The experts take turns asking up to three rounds of questions, and they may guess the organism at any time.
4. Have students take turns playing contestant and expert.

Conclusion

What types of information do you need to identify each organism? (where it lives, what it eats, how it gets its food, what other organisms it coexists with, and so on)

VIDEO NOTES
Habitats of Organisms

Directions Watch the video to match the terms to the correct statement.

1. _____ are living things.

2. An _____ is made of all of the living and nonliving things that surround an organism.

3. A _____ is a place where organisms can live and grow. It's like their home address.

4. A _____ is all the living things in an area.

5. A group of the same kinds of living things in the same area is called a _____ .

6. An _____ is made of groups of living things and their habitats, and it also includes nonliving things.

ecosystem

community

organisms

population

habitat

environment

Choose an organism and describe where it lives by writing about its environment, habitat, ecosystem, community and population. Draw a picture of your organism and where it lives.

Organism

_____

_____

_____

_____

_____

_____

_____

_____

_____

ACTIVITY

Habitats of Organisms

Find Out Who's in my community?

What You Need safety scissors, paper, markers, glue

What to Do

1. Match the organisms that live together in the same community.

2. On separate sheets of paper, draw the habitats where you would find each community of plants and animals.

3. Cut out each picture and paste them into your habitat drawings.

Polar bear

Wildebeest

Fish

Lion

Penguin

Giraffe

Glacier

Serengeti grassy plain

Organisms & Where They Live
Plants & Animals Depend on Each Other

The Big QUESTION *How do plants and animals depend on each other?*

Media Resources
Videotape
Grade 3 • Life Science

Laserdisc
Grade 3 • Life • Side 1

SAMPLE

Frame 12345

1 Introduce

Lesson Background In an ecosystem, different animals and plants play different roles in their competition to survive. Plants are usually producers, meaning they make their own food; consumers cannot make their food and are known by what they eat (herbivores, carnivores, omnivores); decomposers break down the remains of dead organisms. Communities are mutually interdependent groups of populations, a fact most vividly explained in terms of "who eats whom." The predator/prey relationship between some organisms not only satisfies the needs of individual organisms, but also keeps the sizes of populations in balance with their environments.

Activate Prior Knowledge Prompt students to recall their own knowledge of food chains by asking them about the diets of specific animals. *(What does a frog eat? What does a bird eat? What does a snake eat?)* Once they are warmed up, you could shift the conversation to interdependence of organisms by asking, *How does a tree stay alive?* Responses that refer to water, soil and air may be used to talk about decomposers, such as worms, that enrich the soil.

2 Teach

Before Viewing the Video
For Discussion
Ask students to differentiate between predator and prey. Encourage students to add detail by identifying the kinds of animals that are both predator and prey (a snake eats mice, but a snake may be eaten by a large bird) as opposed to those animals that are only predator or prey. Write students' responses in three lists on the chalkboard. Tell students that, as they watch the video, they should look for examples of predators, prey, consumers and producers.

After Viewing the Video
Thinking Critically
Have students recall the animals from the video. Which are prey and which are predators? Revise the list on the chalkboard if appropriate. Next, start two more lists by asking students to name some of the scavengers and decomposers that they saw in the video. Finally, review the concept of consumer versus producer. Which organisms in the video qualify as producers?

Objectives
- give examples of predators and their prey
- describe food chains and food webs
- compare and contrast consumers and producers

Vocabulary
predator, prey, food chain, food web, producers, consumers, scavengers, decomposers

Using Video Notes

Distribute copies of Page 12 to students. Allow time for students to review the illustration and terms on their journal pages. Then, show the video again, and tell students to watch closely for the organisms illustrated on their worksheets.

❸ Close

Checkpoint

- What are some examples of predators and their prey? (Grizzly bears eat salmon; cheetahs eat wildebeests; spiders eat insects.)
- What does a food chain show? (A food chain is the way food energy passes from one organism to another within a community.)
- What does a food web show? (A food web is the linkage of all the different food chains in a community.)
- What is the difference between consumers and producers? (Consumers do not produce their own food and are described by what they eat. Producers are organisms that produce their own food; they are usually plants.)

Activity

What to Expect Students will give examples of how organisms depend upon each other.

Distribute a copy of page 13, long strands of different colors of yarn and a sheet of construction paper to each student. Instruct students to use one color of yarn for each organism that feeds other organisms (for example, the green yarn would connect from the fish to the bird, to the dolphin, to the shark, and to the seal; and the blue yarn would connect from the plankton to the fish and the whale).

Alternate Activity

Find Out What is an ecosystem made of?

What You Need large glass jar, water, soil, plants and seeds, invertebrate animals (earthworms, sow bugs, insects, slugs, snails and so on), cheesecloth, rubber band

What to Do

Design your own ecosystem. What animals and plants would live in the ecosystem of your choosing? Remember, your ecosystem can exist on land or in the water. What will the animals need to eat to survive? How will the plants survive? What conditions will be necessary for you to maintain your ecosystem? How will the animals and plants in your ecosystem interact with each other? Write out a food chain for your ecosystem. Identify producers, consumers and decomposers.

FUN FACT

The archer fish, which ranges from India to the Philippines, Indonesia and Australia, can squirt water about two meters, using its tongue and a groove in the top of its mouth, to knock down insects for food. Because light is refracted through water, the archer fish is able to see well both above and below the water at the same time.

ACROSS THE CURRICULUM

Language Art–Creative Writing
Have students choose an animal and write a story about what it eats. Have them tell their classmates all they can about how the animal gets its food.

VIDEO NOTES

Plants and Animals Depend on Each Other

Directions List each organism below under the correct column in the table. Some organisms will have more than one place in your chart.

Prey	Predator	Scavenger	Decomposer	Producer	Consumer

grass   leaves   bees

bird   giraffe   vulture

banana   insects

cheetah

flowers   worm

wildebeest

ACTIVITY

Plants and Animals Depend on Each Other

Find Out Who eats whom in an ocean food web?

What You Need two different colors of yarn, glue, safety scissors, construction paper

What to Do Cut out the pictures below and connect each organism with pieces of yarn to form a food web that shows who eats whom.

Talk About It Which organisms are producers and which are consumers?

Lesson 2 *Plants and Animals Depend on Each Other*

Media Resources
Videotape
Grade 3 • Life Science

Laserdisc
Grade 3 • Life • Side 1
||| SAMPLE |||
Frame 12345

Objectives
- define adaptation
- relate organisms' adaptations to their needs
- distinguish between inherited and learned characteristics

Vocabulary
behavior, adaptations, camouflage, reproduce, inherited characteristics, learned characteristics

Organisms & Where They Live
Organisms Adapt

The Big QUESTION *How do adaptations allow animals and plants to survive?*

① Introduce

Lesson Background To demonstrate learned versus hereditary behaviors of plants and animals, it is useful to recall different ways in which animals and plants adapt to their habitats. Toads provide a good example: they instinctively camouflage themselves according to their surroundings (they can look just like a log they are sitting on), but they learn that some insects taste bad and are best avoided (the monarch butterfly is very bitter-tasting).

Activate Prior Knowledge To highlight hereditary traits in humans, ask students "trick questions" such as: When did you learn how to yawn? When did you learn how to breathe? When did you learn how to sneeze? Follow up with questions that will elicit learned traits: When did you learn to ride a bike? When did you learn to play checkers?

② Teach

Before Viewing the Video
For Discussion
Reveal the answer to the "trick questions" you just asked: some behaviors are inherited. We don't have to learn inherited traits because we are born with them. Tell students to watch the video closely for learned versus inherited characteristics in plants and animals.

After Viewing the Video
Thinking Critically
Ask students to recall learned and inherited characteristics of different organisms that they saw in the video. Write their responses on the chalkboard in two separate lists. How did organisms use these characteristics to adapt to their environments? What were the types of environments to which the organisms had adapted?

Using Video Notes
Distribute copies of Page 16 to students. Allow students several minutes to review the questions and terms on their worksheet. As they watch the video again, have them fill in the blanks to each question.

3 Close

Checkpoint

- What are adaptations? (Adaptations are the structures or behaviors that help living things stay alive.)
- How do organisms' adaptations help them meet their needs? (Adaptations help living things grow, protect themselves, obtain food, have young and survive in their environment.)
- What are inherited traits? (Inherited traits are characteristics that come from your parents, e.g., height.)
- What are learned behaviors? (Learned behaviors are characteristics that are taught or learned from experience, e.g., riding a bike.)

Activity

What to Expect Students will compare and contrast different habitats and explore why certain animals are found in specific places.

Assign an organism and a "new habitat" destination to each student, e.g., My octopus is leaving its natural habitat in the ocean to visit a new spot in Lake Michigan . Distribute copies of Page 17. As students complete their charts, they will begin to realize that habitat is key to their organism's survival.

Alternate Activity

Find Out Although some animals and plants can adapt to changes, change of locale can be harmful. Various animals and plants can live in places far from their natural habitats, such as aquariums, zoos and botanical gardens, as long as someone takes special care of them.

What You Need a large box, art supplies (markers, crayons, clay), rocks and sticks

What to Do

1. Assign an exotic plant or animal to several groups of students and ask them to design a zoo area in which the plant or animal will live comfortably.

2. Find out what each organism's habitat provides, especially energy and water sources, shelter, and neighboring organisms.

3. Help students construct dioramas of their organism's natural habitat out of art supplies and objects from outdoors.

4. Once each habitat is completed, engage the whole class in organizing an entire zoo.

Conclusion

What are some differences between your local area and the organism's natural habitat?

FUN FACT

Scientists think that honeybees and pigeons navigate by using magnetic tissues inside their bodies as compasses.

ACROSS THE CURRICULUM

Language Arts/Technology
Help students write and conduct an investigative interview with a student who plays the part of an animal or plant. Tell them to find out the organism's source of food, where it lives, who its enemies are, and how it protects itself and its family. If possible, have students videotape their interviews.

VIDEO NOTES
Organisms Adapt

Directions Watch the video. Listen for clues to help you match each term to the correct definition.

1. The body parts of animals and plants are called _____ .

2. How organisms act in their environment is called _____ .

3. _____ are structures or behaviors that help living things stay alive.

4. _____ is coloring, shape and size that helps animals or plants blend in with their environment.

5. _____ is another way organisms adapt. It happens when they have more of their own kind.

> learned behaviors
>
> inherited traits
>
> structures
>
> camouflage
>
> adaptations
>
> behavior
>
> reproduction

6. _____ are adaptations that organisms are born with.

7. Adaptive characteristics that are taught are called _____ .

ACTIVITY
Organisms Adapt

Find Out Why are habitats so important?

What to Do

1. Find out how your organism's natural habitat is different from its new habitat.

My _____ is leaving its natural

habitat in _____ to visit a new

spot in _____ .

	Natural Habitat	New Habitat
Food source		
Water source		
Living space		
Friends		
Enemies		

2. On the list above, circle everything that your organism will need to bring from its natural habitat to the new habitat.

3. List new friends and enemies that your organism will meet, and think of how it could protect itself.

Friends	Enemies	Protection

4. Use the other side of this page to draw or describe adaptations that your organism might develop if it had to live in the new habitat for a long time.

Humans in the Ecosystem
People's Place in the Ecosystem

The Big QUESTION *How do people fit into the ecosystem?*

Media Resources
Videotape
Grade 3 • Life Science

Laserdisc
Grade 3 • Life • Side 1
SAMPLE
Frame 12345

1 Introduce

Lesson Background Encourage students to look at the planet as one ecosystem, because humans inhabit virtually all regions of Earth. Point out to students that some people live in close harmony with the land while others live in industrial centers, far removed from natural environs. Emphasize that everyone still depends upon Earth's resources for survival.

Activate Prior Knowledge Determine student understanding by asking students to name things that humans need to live (water, food, air, shelter). Next, ask students to think of how humans in different parts of the world find and use resources for these requirements.

2 Teach

Before Viewing the Video
For Discussion
Talk to students about the different places that humans live. Responses should be varied, e.g., people live in cities, on farms, in the desert sleeping in a tent and so on. Ask students to name different ways that humans use soil, water and air. Record their responses on the chalkboard. Show the video, and tell students to watch closely to find out more about each of these natural resources.

After Viewing the Video
Thinking Critically
Ask students to name which natural resources are recyclable (air, water) and which are not (soil). How do humans affect these natural resources?

Using Video Notes
Distribute copies of Page 20 to students. Have them watch the video once again. This time, have students illustrate the use and production of natural resources on their worksheets.

Objectives
- describe how people interact with other living things
- identify human needs
- examine how humans fill their needs

Vocabulary
carbon dioxide, natural resources, recycle

③ Close

Checkpoint

- How do people interact with other living things? (As part of the ecosystem, humans not only depend upon other living things, but sometimes we also hurt them.)

- What are human needs? (air, water, shelter, food)

- How do humans fill their needs? (Humans use natural resources to produce food, build shelter, and so on)

Activity

What to Expect Students recognize and record the resources they use in a typical day.

Distribute copies of Page 21. After students complete one day's record, tell them to share their data on group charts, each labeled with a different resource. Follow up by discussing which resources are used the most and which resources could be conserved or recycled.

Alternate Activity

Find Out Are all parts of an ecosystem living?

What to Do

1. On classroom charts or on the chalkboard, write the headings "Living" and "Nonliving."

2. Challenge the students to brainstorm components of your local ecosystem and place them on the charts. ***Note:*** Dead organisms would be classified as "Living." Only inorganic items—minerals, metals, water, air, soil—would be "Nonliving."

3. Start two new charts with the headings "Consumers," "Producers" and "Decomposers" on Chart 1, and "Rocks/Minerals," "Gases" and "Liquids" on Chart 2.

4. Ask for student volunteers to transfer the items from the old charts onto the new charts.

FUN FACT

Scientists have learned to make fuzzy, warm clothes out of recycled plastic drink bottles.

ACROSS THE CURRICULUM

Social Studies—Geography

1. Help students name resources that they use and that come from your state. Examples may include wood for construction and paper, farms that produce food and water sources that provide electrical power and drinking water.

2. Using various sources, including state and local agencies as well as your media center, help your students discover where these resources originate.

3. On a map of your state, indicate by colorcoding and/or symbols where the different resources originate.

VIDEO NOTES
People's Place in the Ecosystem

Directions Watch the video. Then, draw a scene in which people are using soil, water and air. Include everything in the ecosystem that is needed to recycle or conserve each natural resource.

ACTIVITY

People's Place in the Ecosystem

Find Out How many resources do I use in a typical day?

What to Do

1. Starting tomorrow morning when you wake up, start keeping a record of all the resources you use until you go to bed at night. Record each resource in Column 1, and write how you used it in Column 2.

2. After you've completed the first two columns of your chart, find out the source of each resource. Write this information in Column 3.

Resource	Use	Source
air	to blow up a balloon	atmosphere, plants

Talk About It Are any of the resources that you use recyclable? How could you help conserve Earth's resources?

Humans in the Ecosystem
Taking Care of Earth

The Big QUESTION *What can you do to protect Earth's resources?*

Media Resources
Videotape
Grade 3 ● Life Science

Laserdisc
Grade 3 ● Life ● Side 1
SAMPLE
Frame 12345

Objectives
- explain why conservation is important
- describe what happens to the things we throw away
- identify how humans can work to save natural resources

Vocabulary
conservation, environmentalists, biodegradable, compost

❶ Introduce

Lesson Background By making wise choices in the use and conservation of resources and the disposal or recycling of materials, everyone can help reduce pollution and waste in the environment. Give students concrete examples of how humans can contribute to this cause (no dumping in clean water; protect and enrich the land by recycling; help keep the air clean by car pooling).

Activate Prior Knowledge Write "Pollution" and "Waste" at the top of your chalkboard and fill in the columns as students name different kinds of pollution and waste. Next, ask your class to suggest ways to reduce the effects of each environmental offense.

❷ Teach

Before Viewing the Video
For Discussion
Have students think about the things that they throw away every day that they could instead be recycling. Have them watch the video for ways that they could conserve at home.

After Viewing the Video
Thinking Critically
Ask students to name things that they can do at home to conserve natural resources. Can the students think of any ways to improve the recycling efforts in their community? Write their responses on the chalkboard.

Using Video Notes
Distribute copies of Page 24 to students. Give them several minutes to review their journal pages. Show the video once again. This time, have students draw the best conservation destinations for each item of trash named on their worksheet.

❸ Close

Checkpoint

- What happens to the things we throw away? (Garbage is dumped in a landfill where it is covered with dirt and begins to decay. Some garbage is recycled or reused.)

- Why is conservation important? (Conservation decreases the amount of garbage in landfills and helps protect natural resources.)

- How can humans work to save natural resources? (People can help save natural resources by recycling and reusing products, turning off lights and appliances that are not in use, using solar power, and so on.)

Activity

What to Expect Students will become more aware of the amount of litter in their area.

Distribute copies of Page 25 and tell students that you are about to tour some specific areas of the school to look for litter. Tell them to make a tally mark for each piece of litter they find. You may want to do one example together so that it's clear how to fill in the chart. **Safety Note:** Do not let students collect the litter, especially dangerous items like glass and cans.

Alternate Activity

Find Out Students act on ways they can positively influence their community.

What to Do

1. Summarize data from the students' Litter Charts and discuss the need for a litter-reduction campaign at school.

2. Determine the form their campaign will take: news programs, intercom announcements, email, posters, and so on. Discourage the use of too much paper in a campaign to reduce paper litter.

3. Encourage students to use creativity with logos, slogans, upbeat chants, songs or poems.

4. Help the students conduct the campaign for one week.

Conclusion

Did the class make a difference in the amount of litter around the school? Why or why not? Was this a one-time fix, or will they need to continue with reminders? Why or why not?

FUN FACT

One tree can filter up to 60 pounds of pollutants from the air every year. However, every Sunday the United States throws away nearly 90% of the recyclable newspapers, which wastes about 500,000 trees.

ACROSS THE CURRICULUM

Math

Have students measure the amount of litter in your schoolyard before and after the antilitter campaign to determine its effectiveness. Have them plot their data in a chart or graph.

VIDEO NOTES

Taking Care of Earth

Directions Watch the video. Then, draw the steps that garbage would go through in an ideal conservationist's home. Imagine that your trash can contains the materials listed here. Where should each item end up?

aluminum cans
banana peels
junk mail
fish bones
drink bottle
milk container
newspaper
applesauce jar
aluminum foil

ACTIVITY
Taking Care of Earth

Find Out Is littering a problem at our school?

What to Do As you conduct a litter inspection tour, make a tally mark on this chart for each item of litter you find. Write down specific kinds of litter in the "Notes" sections *(drink can, plastic bag)*.

	Cafeteria	Playground	Hallways	Classroom
Paper				
Notes:				
Glass				
Notes:				
Metal				
Notes:				
Plastic				
Notes:				
Other				
Notes:				

Talk About It Which areas are most littered? What are the most common types of litter? Does the school need to "clean up" its act? How could this be done?

LIFE SCIENCE 3

Changes in the Environment
Surviving Changes in the Environment

The Big QUESTION *What happens to plants and animals when there is a sudden change in their environment?*

Media Resources
Videotape
Grade 3 • Life Science

Laserdisc
Grade 3 • Life • Side 1
SAMPLE
Frame 12345

1 Introduce

Lesson Background Change is a natural process and nature has the tendency to restore itself when disrupted. Make sure that students know that natural "disasters" are, as their name suggests, often caused by nature rather than by humans and that while plants and animals may suffer during these events, they sometimes benefit from the changes.

Activate Prior Knowledge Ask students to name different types of abrupt natural changes. (volcano eruption, flood, earthquake, hurricane) If they can recall a specific natural event, prompt students to identify its effects (trees fell down, people lost their houses) and extrapolate by asking them what happened to the animals and plants.

2 Teach

Before Viewing the Video
For Discussion
Ask students to describe what happens in the wintertime when it becomes cold. Tell them that this is a change in the environment that happens every year. Ask students to describe what happens to the plants and animals in their community. Do the leaves fall off the trees? Do some plants die? Are there some animals that students will not see again until the spring? Write some of these changes on the board. Show the video.

After Viewing the Video
Thinking Critically
Ask students to describe the changes that they saw in the video. Did the students notice any disasters that could have been caused by humans? Were any of the changes in the environment preventable? Ask students how they would restore the environment.

Using Video Notes
Distribute copies of Page 28 to students. Give students several minutes to review the tasks assigned on their worksheet. Show the video again and have students create descriptive lists of natural disasters and disasters caused by humans.

Objectives
- investigate what happens to plants and animals after natural disasters
- identify ways that nature restores itself after a natural disaster
- identify ways that humans help nature restore itself after a disaster
- describe how humans affect living things in an environment

Vocabulary
restore

③ Close

Checkpoint

- What happens to plants and animals after natural disasters? (Some plants and animals die, others leave the area, and some survive the disaster. Plants, of course, cannot leave, but their seeds may survive or be carried back into an area by animals.)
- How does nature restore itself after natural disasters? (Certain plants and animals survive, return to the area, and reproduce.)
- How do humans help nature restore itself after a disaster? (Humans plant new plants to restart growth and provide food for animals.)
- How do humans affect living things in an environment? (Humans destroy living things by altering natural environments. For example, streets and buildings cover what may once have been home to many populations of plants and animals. Also, man-made disasters such as forest fires and oil spills are very harmful.)

Activity

What to Expect Students will explore how nature recovers from natural disasters.

Distribute copies of Page 29. Help students choose a type of natural disaster and create a logically ordered booklet of changes that typically follow such an event. Reminding students of the food chain will help them comprehend the logical progression of restoration.

Alternate Activity

Find Out What recent changes have occurred in our area?

What You Need resource books, time in the media center, pens and markers, tape

What to Do

Create a time line of your region for the past 100 years depicting how changes in the environment have affected different organisms' survival. Gather resource materials for the students, including atlases for population data, and schedule time in the media center.

1. Divide students into groups and assign each group a time period of a century.
2. With the help of reference materials, have students record their information on a sheet of paper.
3. Encourage group members to illustrate their time line. Then, put the time periods together to create a class time line.

Conclusion

How do changes in the environment affect the organisms living there? How do population changes, construction and natural disasters affect habitats?

FUN FACT

During periods of extreme drought, the lungfish can live out of water in a state of suspended animation for as long as three years at a time.

ACROSS THE CURRICULUM

Language Arts

Ask students to pretend they are historians recording information about how events such as natural disasters have impacted people in their region. Have them role-play interviewing citizens of the region who have "always lived there," "just moved into the area," "want to move away because of changes," and so on. Remind them to include how the people felt about the region before and after the natural disaster, what it was like to live through the disaster, and how they heard of the event.

VIDEO NOTES

Surviving Changes in the Environment

Directions Watch the video. List each disaster in the correct column. Some disasters may be natural and human-caused. Next, suggest ways that each disaster could be prevented or how nature and humans could help restore the environment.

> oil spill
>
> volcano eruption
>
> earthquake
>
> forest fire
>
> chemical leak
>
> flood

Natural Disasters	Prevention/Restoration

Disasters Caused by Humans	Prevention/Restoration

ACTIVITY

Surviving Changes in the Environment

Find Out How do plants and animals recover after a natural disaster?

What You Need construction paper cut into rectangles (10 cm x 15 cm), crayons or markers, fasteners (yarn, staples or brass paper fasteners)

What to Do

1. Choose one kind of natural disaster and create a sequence book showing how nature recovers. Some things that help nature after a disaster are shown below.

2. Draw and describe at least five things from the list on this page on five pieces of construction paper.

3. Put the five events in the order that they occur in nature, then fasten the pages together to complete your booklet.

Wind blowing

Rain falling

Soil getting richer

Seeds blowing in or being carried in

Seeds starting to grow

Small plants growing

Animals seeking food and shelter

Predators seeking food and shelter

Animals bringing in new plants

Larger trees and bushes growing

Talk About It How does your knowledge of food chains help you construct this booklet? What must happen before animals come back to a damaged area in nature?

Media Resources
Videotape
Grade 3 • Life Science

Laserdisc
Grade 3 • Life • Side 1
||||| **SAMPLE** |||||
Frame 12345

Changes in the Environment
Threats to Survival

The Big QUESTION *Why did some animals and plants become extinct?*

❶ Introduce

Lesson Background Plants and animals become extinct as a result of natural changes or human actions. Emphasize to students that this occurs because such organisms cannot adapt to a particular change in their environment. A good example to use is the extinction of the dinosaurs. Some scientists theorize that a meteorite hit Earth and raised a huge cloud of dust, which blocked the sun, killing the plants that the dinosaurs needed to survive.

Activate Prior Knowledge As a contrast to the extinction of dinosaurs, explain to students the situation of the dodo. The dodo was a bird that lived on an island in the Indian Ocean. It was hunted by soldiers. Dogs, rats and hogs ate its eggs. As the number of adult birds shrank, so did the number of offspring produced. Eventually, the dodo became extinct. Ask students how the dodo might have escaped extinction.

❷ Teach

Before Viewing the Video
For Discussion
Many animals today are on the endangered species list. Ask students to name animals they know to be endangered or threatened. Write their responses on the chalkboard. Do any of the animals named have a local significance? Show the video and encourage students to watch for reasons that plants and animals become extinct.

After Viewing the Video
Thinking Critically
How were the animals on the video endangered? Was it because of human intrusion? Was it because of shrinking habitat? How could humans reverse the trend toward endangerment and extinction?

Using Video Notes
Distribute copies of Page 32 to students. Have them watch the video again. This time, have the students draw the steps of how a fossil is formed on their worksheets.

Objectives
- explain what we can learn from dinosaur fossils
- describe theories explaining how dinosaurs became extinct
- explain why other species become extinct

Vocabulary
fossils, adapt, extinct

3 Close

Checkpoint

- What can we learn from dinosaur fossils? (We can learn about what dinosaurs may have looked like and what they ate.)
- How did dinosaurs become extinct? (No one knows for sure. Scientists think that (a) a meteorite hit Earth and created a dust cloud that killed the dinosaurs' food sources, or (b) a flood killed the dinosaurs, or (c) gradual environmental changes transformed the dinosaurs' swamp habitat into a drier environment to which the dinosaurs could not adapt.)
- Why do other species become extinct? (Animals and plants become extinct due to overhunting, destruction of habitat, or environmental changes to which they cannot adapt.)

Activity

What to Expect Students will focus on beneficial or negative environments, which may help them learn something about their own environment that they may or may not like. This could also be a tool to inspire children to change their environment.

Assign a plant or animal to each student and encourage them to think about their organisms' environments. What do they eat? Where do they sleep? Who are their friends and enemies?

Distribute copies of Page 33. Have students follow the steps outlined to see how a proper environment is crucial to an organism's survival.

Alternate Activity

Find Out What can we do to protect endangered species?

What You Need pencils, construction paper, glue, scissors, magazines

What to Do

Supply students with art supplies and the following instructions:

1. Pick a plant or animal that is on the endangered species list.
2. Draw or cut out a picture of your animal or plant in its natural environment.
3. List three things that you would do to protect this species.
4. Will your species survive? Why or why not?

Conclusion

What needs does this species have?

FUN FACT

In 1880 there were approximately 2 billion passenger pigeons in the United States. Thirty-four years later, in 1914, the species was extinct due to deforestation and overhunting.

ACROSS THE CURRICULUM

Social Studies—Geography
Imagine dinosaurs are alive today. Find places on a map of the world where they would most likely be comfortable. Would they be better suited to desert climates? Glacial? Tropical? Why?

VIDEO NOTES
Threats to Survival

Directions For each animal and its environment, write how the animal is adapted to its environment. Also give examples of animals that would not survive in each of the habitats below.

Polar bear

_____
_____
_____
_____
_____
_____
_____
_____
_____
_____

Fish

_____
_____
_____
_____
_____
_____
_____
_____
_____

Tiger

_____
_____
_____
_____
_____
_____
_____
_____

Giraffe

_____
_____
_____
_____
_____
_____
_____
_____

ACTIVITY
Threats to Survival

Find Out How does an animal or plant survive in its environment?

What To Do Choose either a plant or an animal. In the first box, draw your plant or animal and the environment in which it would thrive. In the second box, draw your plant or animal and the environment in which it would become sick. In the third box, draw your plant or animal and the environment in which it would die.

Changes in the Environment
Survival in a Changing World

The Big QUESTION *Why are some plants and animals endangered?*

Media Resources

Videotape
Grade 3 ● Life Science

Laserdisc
Grade 3 ● Life ● Side 1
SAMPLE
Frame 12345

1 Introduce

Lesson Background Although many environmental changes are natural processes, human actions can greatly impact the survival or extinction of species. For example, the spotted owl was endangered because humans were destroying their natural habitat in forests of the Pacific Northwest. Humans need to be aware of the consequences of their actions to protect our global habitat.

Activate Prior Knowledge Determine student understanding by asking them to name both extinct and endangered species. List these organisms on the board under the labels "Extinct" and "Endangered." To emphasize the concept of global habitat, ask students to name some benefits that different organisms offer to humans. (Rain forests provides oxygen and medicinal plants; animals and plants provide food; all organisms help to maintain balance in their ecosystems.)

2 Teach

Before Viewing the Video
For Discussion
Explain to students that extinction is sometimes a natural occurrence but that humans are hastening extinction for many animals. Some animals (the Blue Whale) have been hunted to near extinction while others have lost their habitat due to human encroachment (the Spotted Owl). Tell students to look for these kinds of examples in the video.

After Viewing the Video
Thinking Critically
What endangered animals were shown in the video? Ask students if they believe that these endangered species will become extinct. What can students do to help prevent these animals and local endangered species from becoming extinct?

Using Video Notes
Distribute copies of Page 36 to students. Allow students to review the terms and statements on their worksheets. Show the video once again. This time, have students match the terms to the statements on their worksheet.

Objectives
- describe how living things change over time
- explain what it means to be endangered
- describe the importance of rain forests
- relate life on Earth to a balance of living things

Vocabulary
endangered

❸ Close

Checkpoint

- How do living things change over time? (Living things change physically over time to adapt to changes in their environment.)
- What does it mean to be endangered? (Living things that are endangered become so few in number that they may not be able to have young. If this happens, their whole population will become extinct.)
- Why are rain forests important to the global ecosystem? (They produce large amounts of oxygen, contain important medicinal plants and provide habitats for many organisms.)
- How does life on Earth depend on a balance of living things? (Every part of our global ecosystem depends on the other parts in some way.)

Activity

What to Expect Students will investigate a particular organism and compose a diary reflecting its daily activities. By simulating an organism's struggle to survive, students will dramatize the importance of balance in individual habitats and ecosystems as a whole.

Distribute copies of Page 37. After students have completed their research, ask them to write their journal entries.

Note: Make extra copies if necessary.

Alternate Activity

Find Out How can human intervention help protect an endangered species?

What You Need poster boards, art supplies

What to Do

1. Have students do research to find one or more examples of endangered animals or plants from your own region.
2. Conduct a class discussion concerning their findings. Ask students: What is being done to help save this plant or animal? What can we do to help?
3. Make a "Save the _____" awareness poster to display at the school or in the community.

Conclusion

Why is this organism endangered? Are there any people who may argue against the measures the class has proposed be taken to protect it?

FUN FACT

Having lived for almost 200 million years, the American alligator was at the brink of extinction and was listed as an endangered species in 1967. From that time the alligator has surged in population growth and has now been declassified as endangered. The alligator's major contribution to their environment is their building of "gator holes," which they dig out from the mud over several years. During the dry season these "gator holes" retain and provide vital water for many organisms in the alligator's habitat.

ACROSS THE CURRICULUM

Social Studies—Geography
Ask students to make a small drawing of an endangered plant or animal. Have them cut out their picture and, using a world map, have them locate the habitat area in which their organism is found. Pin or tape each student's drawing next to the map with a line or piece of yarn pointing to its correct location.

VIDEO NOTES
Survival in a Changing World

Directions Match the terms to the correct definition.

1. The ceiling created by the _____ traps heat which slowly raises Earth's average temperature and creates harmful changes in the global ecosystem.

2. The _____ has been hunted for so long that it is now on the endangered species list.

3. The _____ is endangered because loggers were cutting down the trees in its habitat.

4. In 1973, Congress passed the _____ , which makes it illegal to harm any species on the endangered species list.

5. A plant or animal is _____ when it is in danger of becoming extinct.

6. The _____ produces much of Earth's oxygen, contains important medicinal plants and provides habitats for many different species of plants and animals.

7. Many ecosystems on our planet depend upon each other to make up the _____ , which is Earth's ecosystem.

> **endangered**
>
> **Blue Whale**
>
> **Endangered Species Act**
>
> **Spotted Owl**
>
> **rain forest**
>
> **global ecosystem**
>
> **greenhouse effect**

ACTIVITY

Survival in a Changing World

Find Out How do plants and animals become endangered?

What to Do

1. Choose an endangered animal or plant and research its life: What does my organism look like? Where does my organism live and what does it eat? What is happening to make it endangered?

2. Pretend you are the endangered organism. Write in your diary about your daily struggle to survive.

Date:

_____
_____
_____
_____
_____
_____
_____

Date:

_____
_____
_____
_____
_____
_____
_____

Date:

_____
_____
_____
_____
_____
_____
_____

Date:

_____
_____
_____
_____
_____
_____
_____

Earth and Space
Earth, Sun and Moon

The Big QUESTION *Why does the moon appear to change?*

Media Resources
Videotape
Grade 3 • Earth Science

Laserdisc
Grade 3 • Earth • Side 1
SAMPLE
Frame 12345

Objectives
- investigate and describe the phases of the moon
- relate shadows to opaque objects
- explain what happens during an eclipse

Vocabulary
- shadow, opaque, eclipse, phases

❶ Introduce

Lesson Background The moon travels around Earth while Earth is traveling around the sun. The moon is the brightest object in our sky, but it gives off no light of its own. When it "shines," it is reflecting light from the sun.

From Earth, we see the moon in its different phases. When the moon is between the sun and Earth, the sunlit side of the moon faces away from Earth and we barely see it. Astronomers call this the new moon. Each day, as the moon orbits Earth, we can see more of the sunlit side of the moon. First, we see a small sliver called the crescent moon. In about one week, we see half of a full moon, called the first quarter. When the moon has orbited all the way to the other side of Earth (away from the sun), we see the full moon. As the moon orbits back toward the sun, we see another half of the full moon, called the last quarter, then it wanes to a crescent and back to a new moon at the beginning of another monthly cycle. As the moon changes phases, it appears to change shape; emphasize to students that the moon is actually always the same shape (a globe).

During a solar eclipse, the moon moves into a direct line between Earth and the sun and, because it is opaque, it blocks the sunlight on Earth.

Activate Prior Knowledge Ask students what happens during an eclipse. Write their ideas on the chalkboard. Ask students how Earth and the moon are shaped. Record these ideas as well.

❷ Teach

Before Viewing the Video
For Discussion
Ask a volunteer to be the "sun" by standing still and holding a flashlight. Pick two more volunteers to be "Earth" and the "moon." Establish that the sun is stationary and that the moon and Earth move: first, ask the "moon" to walk around "Earth;" second, ask "Earth" to walk around the sun; finally, coordinate the students so that "Earth" walks slowly around the "sun" while the "moon" walks quickly around "Earth." Demonstrate an eclipse by lining up the volunteers so that the "moon" blocks the light of the "sun," leaving "Earth" in a shadow. After this kinesthetic activity, show the video and ask students to relate their human models to the representations in the video.

After Viewing the Video
Thinking Critically
1. Challenge students to recount all the phases of the moon. (new moon, young crescent, first quarter, full moon, last quarter, old crescent)

2. Ask students what happens during an eclipse. (During a solar eclipse, the moon blocks the sun's light and casts a shadow on Earth; during a lunar eclipse, Earth blocks the sun's light and casts a shadow on the moon.)

Using Video Notes
Distribute copies of Page 40 to students. Have them watch the video again. Tell them to watch closely for information about the phases of the moon and have the students complete the drawings on their worksheets.

❸ Close
Checkpoint
- What are the phases of the moon? (new moon, crescent, first quarter, last quarter, full)
- Why do some objects cast shadows? (When light strikes an opaque object, the object blocks out the light and casts a shadow.)
- What happens during a solar eclipse? (The moon moves between the sun and Earth and, because it is opaque, casts a shadow on Earth.)

Activity
What to Expect Students will simulate a solar eclipse and illustrate the positions of the sun, moon and Earth in a solar eclipse.

Distribute copies of Page 41. Assign students to groups of three members each and have them follow the steps outlined to simulate a solar eclipse.

Conclusions
How long does it take for the moon to pass through one complete cycle of all of its phases? (one month)

FUN FACT
If the moon were placed on the surface of the continental United States, it would extend from San Francisco to Cleveland (3456 kilometers).

Earth moves in its 936-million-kilometer orbit around the sun about eight times faster than the speed of a bullet.

ACROSS THE CURRICULUM
Technology
Listen to a local astronomy program or check an astronomy website *(Earth & Sky has an excellent site)* to determine the time the moon rises each night *(key words: earth, sky, astronomy)*. Does the moon rise at about the same time every night?

VIDEO NOTES
Earth, Sun and Moon

Directions Draw each phase of the moon.

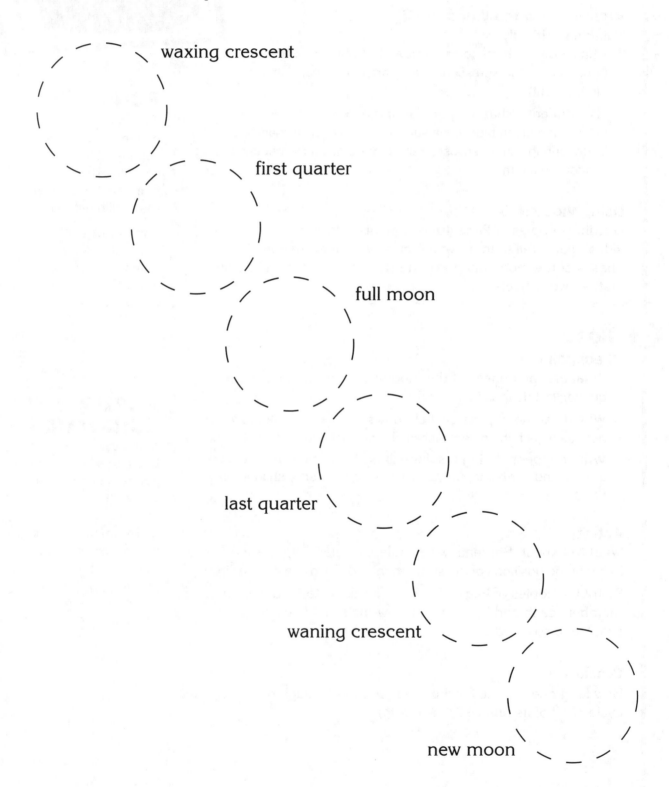

waxing crescent

first quarter

full moon

last quarter

waning crescent

new moon

ACTIVITY
Earth, Sun and Moon

Find Out What happens during a solar eclipse?

What You Need one large and one small polystyrene ball, two sharpened pencils and one flashlight per group

What to Do

1. Working in groups of three, push pencils into polystyrene balls to act as holders. Two students in each group should hold the balls while the third student holds a flashlight. **Note:** The larger ball represents the Earth, the smaller ball represents the moon and the flashlight represents the sun.

2. Hold the "moon" between the "sun" and "Earth." Move the "moon" around "Earth" to show its orbit.

3. Stop the "moon" at the place where it blocks the "sun" and casts a shadow on "Earth," simulating a solar eclipse.

4. Move the "moon" closer to or further from the "Earth" until the entire "Earth" disc is in shadow.

5. Draw and label a diagram below, showing the positions of the sun, moon and Earth in a solar eclipse.

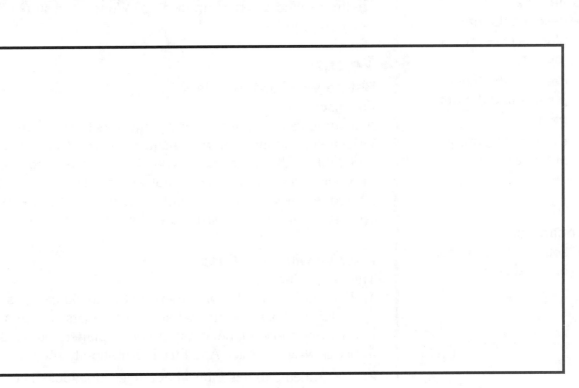

EARTH SCIENCE
1

Earth and Space
The Nine Planets

The Big QUESTION *What makes up our solar system?*

Media Resources
Videotape
Grade 3 • Earth Science

Laserdisc
Grade 3 • Earth • Side 1
SAMPLE
Frame 12345

Objectives
- describe the path planets take around the sun
- describe the inner planets and the outer planets
- explain the purpose of a telescope

Vocabulary
- solar system, planet, orbit, telescope

1 Introduce

Lesson Background At the center of the solar system is a star we call the sun. The huge mass of the sun creates gravitation that keeps the nine planets and their satellites (or moons) moving in elliptical orbits around the sun. The planets have been classified as either inner planets or outer planets. The inner planets (Mercury, Venus, Earth, Mars) are smaller, warmer, and composed of rock while most of the outer planets (Jupiter, Saturn, Uranus, Neptune, Pluto) are larger, colder and have a gaseous composition. Pluto is classified as an outer planet because of its location, but it is small and composed of rock. The solar system also contains asteroids, meteoroids and comets.

The telescope is an instrument used to observe distant objects by magnifying them. Hans Lippershey, a Dutch optician, probably made the first telescope in 1608. A year later, Galileo built his own telescope and began making important discoveries, including four of Jupiter's satellites, the rings of Saturn, and mountains and craters on the moon. Most importantly, Galileo was able to confirm earlier theories that the sun is the center of the solar system.

Activate Prior Knowledge Ask students basic questions about the solar system. What is in the solar system? How many planets are there? Where is the sun located? What is the sun? What is a telescope used for?

2 Teach
Before Viewing the Video
For Discussion
Ask students to name as many planets as they can remember. Write their responses on the chalkboard. Then, ask students to name things that we can see without using a telescope (some stars, a few planets, the sun, the moon). What would we see if we did use a telescope? (more detail, more planets, more stars) Tell students to watch and listen carefully to learn more about our solar system.

After Viewing the Video
Thinking Critically
1. Working from the list of planets on the chalkboard, ask students to identify those planets that are inner planets (Mercury, Venus, Earth, Mars) and those planets that are outer planets (Jupiter, Saturn, Uranus, Neptune, Pluto). Add this information to the list.

2. Ask students to describe what a telescope looks like. (a tube with disks of special kinds of glass called lenses)

3 **Using Video Notes**

Distribute copies of Page 44 to students. Give students several minutes to review the questions and respond to any they can. Then, show the video once again. Have students answer the questions on their worksheet.

Close

Checkpoint

- What kind of path do planets follow as they travel around the sun? (an elliptical orbit)

- Describe the inner planets. (closer to the sun, warmer, smaller, mostly made up of rock)

- Describe the outer planets. (further away from the sun, colder, all larger and gaseous except for Pluto, which is small and solid)

- What does a telescope do? (makes distant objects look bigger and closer)

Activity

What to Expect Students will describe the composition of each planet.

Distribute copies of Page 45. Have students use research books and the Internet to complete their charts.

Alternate Activity

Find Out What makes up a solar system?

What to Do

1. Using a diagram and the facts of our solar system as a guide, have students invent and draw their own solar systems, naming the planets and sun(s).

2. Tell them to draw one or two central stars and positon at least four planets revolving around the star(s). The students should also decide whether or not their planets have moons.

3. Have students describe the composition of each planet. Is it gaseous or solid? Is it cold or hot? Is it big or small? Is there liquid water on the planet? Students should draw and color each planet according to its composition.

Conclusion

What if telescopes allowed us to observe other stars in great detail? Do you think we would find planets orbiting them?

FUN FACT

The word planet comes from the Greek word, *planetes*, which means wanderer.

ACROSS THE CURRICULUM

Language Arts

Have students create a travel brochure for each of the planets. Bring in samples from a travel agency to give them ideas on what to include. Tell students the brochures must contain accurate information and directions on how to get to the planet. Display all the brochures.

VIDEO NOTES

The Nine Planets

Directions Fill in the blanks or write a short answer.

1. A telescope makes distant objects appear _____.

2. An orbit is _____.

3. Planets travel in an orbit around _____.

4. The sun is located _____ of the solar system.

5. List the inner planets: _____ _____

_____ _____

6. List the outer planets: _____ _____

_____ _____

_____

These questions might take a little more thought ...

1. Could you survive on any other planet in the solar system? Why or why not?

_____

_____

_____

_____

2. If you had a spaceship and wanted to travel to another planet, which one would you pick? Why? On which planets would you be able to land your spaceship? _____

_____

_____

_____

3. Which planet do you think is most different from Earth? Explain your choice. _____

_____

_____

_____

ACTIVITY
The Nine Planets

Planet	Air	Water	Land	Moons
1._____				
2._____				
3._Earth_____	Yes	Yes	Yes	One
4._____				
5._____				
6._____				
7._____				
8._____				
9._____				

Earth and Space
Constellations

The Big QUESTION *Why do constellations appear to change with the seasons?*

Media Resources
Videotape
Grade 3 • Earth Science

Laserdisc
Grade 3 • Earth • Side 1
SAMPLE
Frame 12345

Objectives
- describe the Milky Way and the Big Dipper
- explain the uses of a star chart

Vocabulary
- galaxy, stars, constellations, planetarium

① Introduce

Lesson Background The universe as we know it consists of an infinite number of galaxies similar to our own, the Milky Way, which consists of hundreds of billions of stars, including the sun. The sun is a star—an enormous ball of burning gases formed from huge, spinning clouds of gas and dust. The sun is at the center of our solar system, and everything in our solar system orbits around it, held by the sun's gravitational force. Our solar system consists of the nine known planets and their moons, plus asteroids, meteoroids and comets.

A constellation is a group of stars that appear together and resemble an image of a person, animal or object. The constellations that we see today are the same ones that the Greeks observed and named 2000 years ago. Because Earth rotates, the stars in the constellations appear to "move" each night. Depending on your location, the stars appear to rise and set, just like the sun. As Earth orbits the sun, some constellations gradually move out of sight and new ones come into view, following a yearly pattern. This is why we have summer and winter constellations.

Activate Prior Knowledge Ask students if they have ever taken a trip to a planetarium (a museum in a dome-shaped room with the night sky projected on the ceiling). What did they see? Or, ask students to name the things they see when they look up into the night sky.

② Teach

Before Viewing the Video
For Discussion
Write the words *moon, Earth, Milky Way galaxy* and *solar system* on the chalkboard in random order. Ask students to tell which is the largest and which is the smallest.

After Viewing the Video
Thinking Critically
1. With feedback from students, reorder the words on the chalkboard so that they are in ascending order according to size.

2. Ask students to name their favorite constellations. Which ones can they observe year-round? (the Big Dipper) Why do the constellations seem to move? (because the Earth is rotating)

Using Video Notes

Distribute copies of Page 48 to students. Have them watch the video again and match the correct word to its definition on their worksheets.

❸ Close

Checkpoint

- What is the Milky Way? (our galaxy, a family of hundreds of billions of stars)
- What is the Big Dipper? (a constellation, resembling a water ladle, that can be seen year-round from the northern hemisphere)
- What is a star chart? (a map of the sky used to locate constellations during different seasons)

Activity

What to Expect Students will illustrate constellations and simulate the appearance of star movement.

Distribute copies of Page 49. Have students follow the steps outlined to see how constellations are grouped and how they appear to move. Explain to the students that this activity approximates what they see in the sky as they watch the stars. Stress the fact that the stars seem to move in the sky only because Earth (spinning chair) is rotating as it revolves around the sun.

Alternate Activity

Find Out How do the appearances of constellations change?
What to Do
1. Have students, with the help of an adult, look at the night sky and locate the Big Dipper and another constellation.
2. Make a drawing of the constellations' exact locations by including the location of a tree or other ground object in the picture.
3. One hour later, observe and draw any changes in the positions of the stars.

Conclusions

What happened to the constellation after an hour? (It moved across the sky.) Explain why this happens. (Earth is rotating.)

Do you think that you would find the constellation in the same place six months from now? (No. As Earth orbits the sun, some constellations gradually move out of sight and new ones come into view, following a yearly pattern.)

FUN FACT

The fastest jet would take a million years to fly to the nearest star past the sun. That's a one-way trip of more than 40 trillion kilometers.

ACROSS THE CURRICULUM

Language Arts
Have each student pick a constellation and research its mythology. Students should also draw their constellations and the figures associated with them. Have students report their findings to the class.

VIDEO NOTES

Constellations

Directions Watch and listen for clues to the definitions. Write the correct word beside its definition.

1. _____ A group of billions of stars, gas and dust held together by gravity, to which our sun belongs

2. _____ A museum for viewing stars, constellations and planets

3. _____ A map to help you locate constellations at different seasons

4. _____ Enormous balls of burning gases formed from huge spinning clouds of gas and dust

5. _____ Groups of stars that resemble different images

6. _____ A constellation that looks like a water ladle and can always be seen from the northern hemisphere

Bonus Question:

_____ A group of planets held by gravity to a central star; a sun system

constellations

star chart

planetarium

the Milky Way galaxy

stars

the Big Dipper

solar system

ACTIVITY
Constellations

Find Out Why do the stars seem to move?

What You Need scissors, black construction paper, glue, flashlight

What to Do

1. Glue the constellation sheet onto a piece of dark construction paper. Cut the pattern along the dotted line.

2. Use a sharp pencil point or pen tip to punch out the stars. Roll the sheet to form a cone and glue the two ends together.

3. Sit in a swivel chair. Ask another student to hold the cone above you and shine the flashlight into the cone to project the star pattern onto the walls and ceiling of a darkened room. Slowly turn in your chair, which represents Earth. Watch the stars "move."

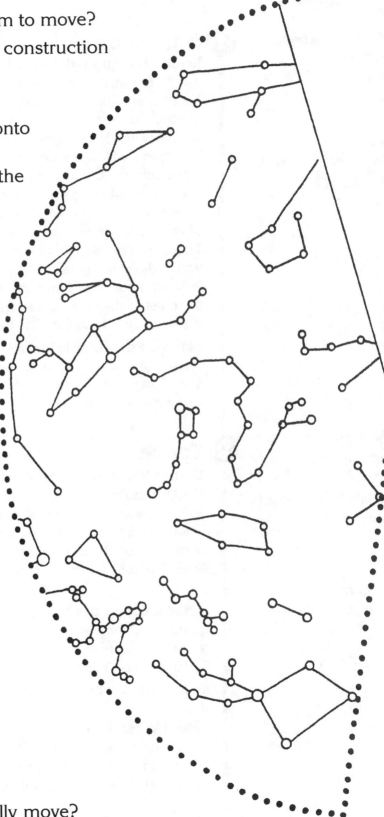

Talk About It Do the stars really move?

EARTH SCIENCE
2

Earth and Its Many Layers
Earth's Composition

The Big QUESTION *What is inside Earth?*

Media Resources
Videotape
Grade 3 • Earth Science

Laserdisc
Grade 3 • Earth • Side 1
SAMPLE
Frame 12345

Objectives
- describe Earth's layers
- explain how scientists study Earth

Vocabulary
crust, mantle, core

1 Introduce

Lesson Background By studying the records of earthquake waves, scientists have found out that Earth's interior is made of hot rock and metal. The three basic layers of Earth's composition are the crust, the mantle and the core. The mantle is divided into an upper and lower layer, and its upper layer is further divided into an upper and lower zone. The upper zone is hard and rigid and, along with the crust, consists of 15 plates that move on top of the lower zone, which is soft molten rock. This movement, continental drift, has resulted in the separation of continents and the formation of various landforms, such as some mountains (the Himalayas). The core is divided into an inner and outer core. The inner core is iron and nickel made solid by immense pressure. The outer core is liquid metal. The temperature of Earth's layers increases from the crust to the core, where temperatures range from 2200° C in the outer core to 5000° C in the deepest parts of the inner core.

Activate Prior Knowledge Ask students to describe the solar system. (the sun and the nine planets around it) How is the sun different from Earth? (The sun is a star—a ball of burning gases; Earth is a planet that is cool on the surface, forming many different landforms.)

2 Teach

Before Viewing the Video
For Discussion
Ask students to imagine what a peach or an avocado looks like when it is sliced in half. As they describe each layer, write their responses on the chalkboard. Then, tell students to watch and listen carefully for more information about Earth's layers.

After Viewing the Video
Thinking Critically
1. Ask students to name the layers of Earth they can recall. (crust, upper mantle, lower mantle, core)
2. Is Earth a solid? (No. Some layers are solid but others are liquid.)

Using Video Notes
Distribute copies of Page 52 to students. Play the video once more and have students determine whether each statement on their worksheet is true or false.

③ Close

Checkpoint

- What are the layers of Earth called? (core, mantle, crust)
- How do scientists find out what is inside Earth? (by studying earthquake waves)

Activity

What to Expect Students will illustrate Earth's layers by drawing or constructing a scale model of Earth's interior.

Distribute copies of Page 53 and have students color-code a drawing, or model from clay, their own cross-section models of Earth's interior.

Alternate Activity

Find Out What are the common misconceptions about the composition of Earth?

What You Need students and adults to interview, paper, pencil

What to Do

Explain to students that many people do not typically remember what they may have learned about the layers of Earth. In order to find out what people think Earth is made of, have students conduct the following social experiment.

1. Write a brief survey of questions targeting people's concepts of Earth's layers. Ask questions to find out what people typically think Earth is like at its core and between the crust and the core.

2. Conduct a survey of fellow students and adults.

3. Organize your findings in a chart. How many people think that Earth is solid? How many people know about the soft layers?

Conclusion

What common misconceptions about Earth's composition did you uncover?

ACROSS THE CURRICULUM

Math

Have students use the depths of each layer of Earth (see Activity) to figure out the diameter of Earth (about 12,670 km).

VIDEO NOTES

Earth's Composition

Directions Read the following statements carefully and decide whether they are true or false. Watch the video for clues to the answers.

T *or* F	1.	Earth's core is a solid ball made of nickel and iron.
T *or* F	2.	Earth's crust is divided into an upper zone and a lower zone.
T *or* F	3.	The plates in Earth's upper zone move over the soft lower zone.
T *or* F	4.	The top layer of Earth is called the crust.
T *or* F	5.	Fossils can be found in Earth's crust.
T *or* F	6.	The mountains that we see on Earth have always been there.
T *or* F	7.	The crust is thinner over land and thicker under oceans.
T *or* F	8.	Scientists learn about Earth's different layers by studying earthquake waves.
T *or* F	9.	The mantle is the middle layer of Earth, between the crust and the core.
T *or* F	10.	Earth below our feet is constantly changing.

ACTIVITY
Earth's Composition

Find Out What is Earth made of?

What You Need markers, paper, modeling clay (optional)

What to Do

Draw or shape a scale model of Earth's composition using the information provided below. Color-code each layer of your model to show an accurate cross section.

Inner Core 1200 km	solid ball of iron and nickel
Outer Core 2200 km	liquid metal
Mantle 2900 km	liquid rock
Crust 35 km	rigid and hard

Talk About It What feature of the mantle allows the plates in the crust to move?

Earth and Its Many Layers
Earth's Forces

The Big QUESTION *What can cause landforms to change?*

Media Resources
Videotape
Grade 3 • Earth Science

Laserdisc
Grade 3 • Earth • Side 1
SAMPLE
Frame 12345

Objectives
- name different forces that change Earth's surface
- explain how some landforms change very slowly
- explain how some landforms change very quickly

Vocabulary
landform, deposition, glacier, volcanoes, earthquake, fault

1 Introduce

Lesson Background Earth's surface is distinguished by many different landforms, such as mountains, valleys, plains, rivers, lakes, islands and beaches. These landforms are continually changing, even though some are changing at an almost imperceptibly slow pace. Emphasize to students that although we cannot see some changes as they are happening, scientists are able to take measurements that prove that landforms are on the move. For example, glaciers move slowly (at a top speed of 30 cm per day), but they produce massive results (the Great Lakes). A nice analogy to give students concerns the hour hand of a clock: it seems to remain still if watched for five minutes, but after an hour its movement is quite obvious. Conversely, some changes happen very quickly and even endanger lives (volcanic eruptions, earthquakes, mudslides).

Activate Prior Knowledge Ask students to name land features on Earth's surface (mountains, valleys, lakes, islands, beaches, rivers). Write the word "Landforms" at the top of the chalkboard and add students' responses to form a list.

2 Teach

Before Viewing the Video
For Discussion
Ask students to name specific landforms in your area. Add these surface features to the list. Ask students to give examples of what can make landforms change. Record their responses on the chalkboard. Show the video.

After Viewing the Video
Thinking Critically
Have students name some of the natural forces that change landforms (gravity, moving water, weather, earthquakes, volcanoes, ice, water, wind, mudslides, glaciers, waves on the beach, heavy storms and floods). Add these forces, where appropriate, to the list on the chalkboard.

Using Video Notes
Distribute copies of Page 56 to students. Play the video again, and tell students to watch closely for fast and slow changes to Earth's landforms. Have them list these changes on their worksheets.

③ Close

Checkpoint

- What are some forces that change Earth's surface? (gravity, volcanoes, earthquakes, glaciers, weather and climate changes, deposition, water)
- What forces change landforms very slowly? (glacier movement, wind and rain)
- What forces change landforms very quickly? (volcanoes, earthquakes, rockslides, heavy storms and floods)

Activity

What to Expect Students will identify forces that created some of Earth's landforms.

Distribute copies of Page 57 and instruct students to explain how each landform was created.

Alternate Activity

Find Out How does water shape the surface features of Earth?

What You Need sand, small gravel, dirt, plastic tub or deep tray, two paper cups, water, pencil, two books or blocks of wood

What to Do

1. Assign students to groups and have them layer the gravel, sand and dirt at one end of the tray to model land. Smooth its surface. Leave the other end empty to collect water.

2. Slowly drip water over the land until it is wet.

3. Place a book under the end of the tray with the land.

4. Use just the tip of a pencil to poke a very tiny hole in the bottom of one paper cup.

5. Hold the cup with the hole above the land. Fill it with water using the other cup. Watch what is happening to both the water and the land while the water drains.

6. After the water has stopped flowing, draw a picture of the land's new form and record your observations.

Conclusion

How did the "land" change? (Part of the soil and sand and some of the gravel washed to the other end of the pan, leaving a valley.)

What caused these changes? (The moving water pushed and carried the soil, sand and gravel down the pile as it rushed to a lower level.)

How could water change rocks and land on Earth? (Water can move rocks and soil, changing the shape of landforms.)

VIDEO NOTES

Earth's Forces

Directions Watch the video carefully. List Earth forces that change landforms quickly or slowly.

Fast changes	Slow changes

Can you think of any changes that take a medium amount of time? _____

ACTIVITY

Earth's Forces

Find Out How on Earth did this happen?

What to Do Tell how you think each landform has been changed over time.

Sand dunes

Ocean arch

Land arch

Talk About It How are the land arch and the ocean arch different?

Earth and Its Many Layers
Surface Features of the Earth

The Big QUESTION *How are Earth's surface features formed?*

Media Resources
Videotape
Grade 3 • Earth Science

Laserdisc
Grade 3 • Earth • Side 1
SAMPLE
Frame 12345

Objectives
- name some different landforms
- explain how a topographical map shows Earth's features in a special way

Vocabulary
mountain, plain, topographic maps, plateau

1 Introduce

Lesson Background This lesson centers on three specific types of landforms: plateaus, plains and mountains. Since students learned about the movement of crust plates in Lesson One, you may want to inform them that these movements are one of the very powerful "forces inside the Earth" that create mountains. Glaciers, volcanoes and water, including rivers, lakes and oceans, form plains. Each of these forces leaves behind a specific type of plain, namely, glacial plains, lava plains, river plains, lake plains and coastal plains. Plateaus, also called tablelands, form when flat rocks are pushed up by forces under the ground. Plateaus do not crumble and fold, as mountains do, but remain flat on top. Canyons form when rivers cut through plains and plateaus.

Topographic maps aid navigation by not only displaying area and distance, but also by providing information about landforms. Contour lines on a topographic map indicate elevation in relation to sea level.

Activate Prior Knowledge Ask students if they have ever been on a vacation to, or seen pictures of, a popular tourist park such as the Grand Canyon, the Smoky Mountains, the Grand Tetons, Monument Valley, etc. What did the landforms look like?

2 Teach

Before Viewing the Video
For Discussion
Draw three columns on the chalkboard with the titles "Mountains," "Plateaus" and "Plains." Ask students to recall whether they saw any of these landforms during their vacations. Write the locations under the appropriate columns. Review the list on the chalkboard and ask students how they think each landform might have been formed. Write their responses beside each location, then show the video.

After Viewing the Video
Thinking Critically
1. Have students review their list of landforms and their ideas on how each was formed. Based on the information in the video, do they want to change their minds?
2. Ask students to imagine that they are going for a hike in the mountains. How would a topographic map help them find the best path? (Contour lines that indicate elevation would help them pick the easiest path or help them find peaks and valleys.)

Using Video Notes

Distribute copies of Page 60 to students. Have them watch the video again, looking for clues that will help them match the correct words to the definitions and Earth forces described on their worksheet.

③ Close

Checkpoint

- What are the three general landforms on Earth? (mountains, plateaus, plains)
- What special information does a topographic map provide? (Contour lines indicate elevation.)

Activity

What to Expect Students will examine a topographic map and chart two routes of different elevations through a mountainous area. Distribute copies of Page 61 and have students locate the lowest path and the highest path.

Alternate Activity

Find Out What landforms can be found in your state?

What You Need a map showing landforms in your state and in the United States

What to Do

1. Have students find different types of landforms in your state and divide the state into regions based on these landforms.

2. Next, have students divide the United States into regions by landforms.

Conclusion

In which landform region(s) is your state located?

FUN FACT

The Himalayan mountain range, which contains the highest mountains in the world, was formed when the Indo-Australian continental plate slowly crashed into the Eurasian plate millions of years ago.

ACROSS THE CURRICULUM

Social Studies

Ask students to work in groups and use a world map or a globe to find three mountain ranges in the United States and two mountain ranges in other countries. Have students reference an encyclopaedia or the Internet to find out more about each mountain range. What continent is the range or peak on? What does it look like? How was it formed? Ask a volunteer from each group to report his or her group's findings to the class.

VIDEO NOTES
Surface Features of the Earth

Directions Watch and listen for clues to the definitions. Write the correct word beside its Earth force.

_____ 1. Formed when flat rocks are pushed up by forces under the ground

_____ 2. Formed by great forces inside the Earth

_____ 3. Formed when a glacier erodes away and leaves behind soil and rock matter it has gathered in its path

_____ 4. Formed as the Colorado River flowed across the plain and slowly cut through the Earth

_____ 5. A map that shows elevation of landforms

_____ 6. Special map lines that connect points of equal height or elevation

_____ 7. Formed from lava spewing over the mountainsides and then cooling

lava plain

topographic map

glacial plain

mountain

plateau

contour lines

the Grand Canyon

ACTIVITY
Surface Features of the Earth

Find Out How useful is a topographic map?

What to Do Using the topographic map as your guide, plan two hikes from point A to point B.

1. Chart one high elevation path that requires climbing from peak to peak.
2. Chart one low elevation path that winds through the valleys.

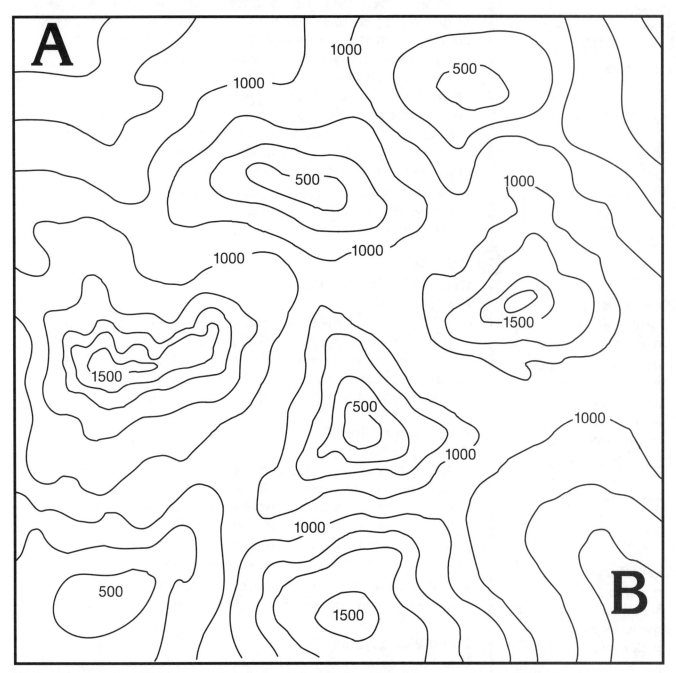

Talk About It Which path do you think would take longer to hike?

Earth Beneath You
Rocks

The Big QUESTION *How do rocks form?*

Media Resources
Videotape
Grade 3 • Earth Science

Laserdisc
Grade 3 • Earth • Side 1
SAMPLE
Frame 12345

Objectives
- identify what rocks are made of
- compare and explain the processes that form igneous, sedimentary and metamorphic rocks
- trace paths through the rock cycle and explain each change

Vocabulary
minerals, igneous, sedimentary, metamorphic, rock cycle

① Introduce

Lesson Background The rock cycle is a model of the slow but continuous changing of rocks from one type to another. Igneous rocks are formed by magma that is either forced to the surface as lava (smoother mineral grains such as pumice or obsidian) or cools slowly beneath Earth's surface (coarser mineral grains such as granite). These rocks rise to the surface where they are broken down by weathering and erosion. The resulting rock particles, along with plant and animal remains, form sedimentary rock (coal, limestone, sandstone, chalk). Most sedimentary rocks form in water. As layers settle on other layers, the layers beneath are compressed and become rock. Metamorphic rocks form when existing rocks are exposed to tremendous heat and pressure beneath Earth's surface (marble, quartz, slate). When this happens, the materials in the rock can undergo both physical and chemical changes.

Activate Prior Knowledge Ask students to name some of the things we find in Earth's crust (minerals, rocks, soil, fossils). Ask students if they think all rocks are made the same way.

② Teach
Before Viewing the Video
For Discussion
Write the names of the three types of rock on the chalkboard with "Igneous" at the upper right, "Sedimentary" at the upper left and "Metamorphic" at the bottom. List specific kinds of each rock beside each type. Tell students to watch the video for clues about how each type of rock is related in the rock cycle.

After Viewing the Video
Thinking Critically
1. Ask students where marble comes from (limestone that was under great pressure and heated). What is the direction of the rock cycle? (Igneous rock is weathered and erodes to contribute to the formation of sedimentary rock while rock beneath the ground is pressured and heated into metamorphic rock, some of which melts and is pushed to the surface to form igneous rock.)

2. Ask students to name some uses for minerals and rocks (jewelry, vitamins, material for buildings, statues).

Using Video Notes
Distribute copies of Page 64 to students. Play the video once again and tell students to watch for clues to the answers on their worksheet.

3 Close
Checkpoint
- What are rocks made of? (Igneous rocks are made of magma and lava; sedimentary rocks are layered pieces of rocks and once living plants and animals; metamorphic rock is formed from other types of rocks.)
- How do igneous, sedimentary and metamorphic rocks form? (Igneous rock forms when lava is pushed to the surface and cools; sedimentary rock forms when pieces of rocks and once living plants and animals settle in layers in water; metamorphic rock forms when heat and pressure change rocks from one form into another.)
- What is the rock cycle? (The constant changing that happens to rocks: igneous rock is beaten down by weather and erodes; pieces of these rocks sink in layers with once living plants and animals and are pressed down to form sedimentary rock; rock beneath the ground is changed by pressure and heat into denser and harder metamorphic rock; minerals and rocks beneath the surface melt into liquid form and are forced to the surface as lava, which cools to form igneous rock.)

Activity
What to Expect Students will make a model of a sedimentary rock.

Distribute copies of Page 65. Have students use a hand lens to examine the sand before and after it has been cemented. To sharpen their observations, ask them to draw what they see, paying special attention to the spaces between the sand and gravel grains.

Note: Corn syrup may be substituted for the sugar solution. The "rocks" may stick to the sides of the cup; have students work patiently and slowly peel away the cup.

Alternate Activity
Find Out How many different kinds of rocks are in my neighborhood?

What You Need local rocks, egg cartons, adhesive labels

What to Do
Encourage students to start a rock collection by collecting rocks in the local area. Students should look for igneous, metamorphic and sedimentary rocks and they should organize rocks by type. Suggest that students reuse egg cartons, as they make excellent containers for small rocks and minerals.

Conclusion
Are all three types of rocks available in your area? What do the local rocks tell you about the Earth history of your area?

FUN FACT
The oldest rocks in the world were found 320 km north of Yellowknife, Canada and are called the Acasta Gneisses. Found in May 1984, these rocks have been dated to 3,962 million years.

ACROSS THE CURRICULUM
Art & Language Arts
Have students bring a rock to school to create a "pet rock." Provide paint, construction paper and glue for them to design their pet rocks. Tell students to give their rocks a name that relates to one of its characteristics, such as how it was formed or what it is used for, and have them write a story about it.

VIDEO NOTES

Rocks

Directions Watch the video and complete the sentences using the terms below.

1. The change of one kind of rock to another is a part of the _____ .

2. Rocks that form when lava cools are called _____ rock.

3. Heat and pressure change rocks into _____ rock.

4. _____ are made of nonliving materials called elements.

5. When small pieces of once living plants and animals get washed into lakes and oceans, these bits and pieces sink in layers to form _____ rock.

> metamorphic
>
> sedimentary
>
> rock cycle
>
> igneous
>
> minerals

Directions Write the letter of the correct answer in the blank.

_____ 6. Because heat and pressure in Earth's crust change limestone to marble, marble must be a(n) _____ .

     a. weathered rock   b. igneous rock

     c. metamorphic rock   d. sedimentary rock

_____ 7. For an igneous rock to change into a sedimentary rock, the igneous rock must _____ .

     a. melt   b. weather and erode

     c. be buried   d. be heated under pressure

_____ 8. The rock cycle refers to _____ .

     a. changes in rocks due to weathering, erosion, melting, cooling, heat and pressure

     b. smaller rocks changing into larger rocks

     c. the rolling of rocks down a hill

     d. all of the above

ACTIVITY

Rocks

Find Out How are sedimentary rocks formed?

What You Need two small paper cups, spoon, sand, small gravel, sugar, water, waxed paper, hand lens

What to Do Pour a spoonful each of sand and gravel in a cup. Fill another cup with 1 cm of water. Add 5 spoonfuls of sugar and stir until it is all dissolved. Slowly pour sugar water into the sand and gravel until it is moistened. Pour off extra water. Let it dry. Over a piece of waxed paper, carefully tear the cup off. Let the "rock" harden for 2 days.

1. Draw a picture of the grains of sand and gravel before they were cemented together.

2. Now draw a picture of the grains of sand and gravel after they were cemented together.

3. What kind of rock did you make? _____

4. How is your "rock" like a real rock? _____

_____

5. How is it different? _____

_____

Earth Beneath You
Soils

The Big QUESTION *How do soils differ from place to place?*

Media Resources
Videotape
Grade 3 • Earth Science

Laserdisc
Grade 3 • Earth • Side 1
SAMPLE
Frame 12345

Objectives
- describe how soil is formed
- describe the three main types of soil
- explain why soil is important

Vocabulary
soil, humus, clay soil, sandy soil, loam

1 Introduce

Lesson Background Soil—a mixture of rock, organic material, water and air found on Earth's surface—is the final product of weathering; it forms only after rock has been broken up into small particles and minerals have been released through chemical weathering. Plant and animal remains decay and form humus, which mixes with rocks and minerals to enrich the soil's nutrients. Plants and animals also aid soil development by tunneling through the soil, making it more permeable. Plant roots even grow into cracks of rocks, wedge them apart and assist in the process of weathering.

There are thousands of kinds of soil on Earth. Soils can be black, red, brown, gray, yellow and combinations of these colors. The soil's color depends on its chemical composition. The three main types of soil are clay, sand and loam. Clay is nonporous and holds water, whereas sand quickly drains water. Loam is a mixture of clay, sand and humus; it is the best soil for growing plants. Only certain plants can grow in either clay or sand.

Activate Prior Knowledge Ask students what they think soil consists of. (rocks, sand, clay, humus, air, water, living and dead things) Record their responses on the chalkboard. What do we need soil for? (growing plants)

2 Teach
Before Viewing the Video
For Discussion
Expand on the question of why humans need soil. Ask students to identify things we depend on that come directly or indirectly from soil. (plants that we eat, plants that animals eat, materials to build houses and so on) Tell students that as they watch the video they should look for reasons why people need soil.

After Viewing the Video
Thinking Critically
1. Ask students to discuss the steps of how soil is formed. (Dead plants and animals decay to form humus; tiny bits of broken rock combine with humus; plants and animals burrow through materials, adding air pockets; rain provides water, and so on.)
2. Ask students to name the three types of soil. (sand, clay, loam) Write these on the chalkboard. Have students specify adjectives that correctly describe each type of soil. Write these responses on the chalkboard under each header.

Using Video Notes

Distribute copies of Page 68 to students. Have them watch the video again, looking and listening closely for answers to the questions on their worksheet.

❸ Close

Checkpoint

- How is soil formed? (When organisms die and decompose, organic material is formed, which returns nutrients to the soil. This organic material is called humus. Humus mixes with rock particles and forms soil.)
- What are the three main types of soil? (clay, sand, loam)
- Why is soil important? (It supports life that we depend upon for food and materials.)

Activity

What to Expect Students will describe soil from an animal's point of view, recognizing the interactions between soil, animals and plants and the importance of soil.

Distribute copies of Page 69. Ask students to choose one earth-tunneling animal and answer the questions on their activity sheets.

Alternate Activity

Find Out How do plants aid soil development?

What You Need lima, kidney, or lentil bean seeds, water, two 35-mm film containers, plastic wrap, two rubber bands

What to Do

1. Fill both containers with as many bean seeds as they'll hold, then fill one container to the top with water. Cover both containers with plastic wrap and secure with rubber bands.

2. Observe the containers the next day.

Conclusion

What happened to the containers of bean seeds? (The bean seeds with no water stayed the same, while the bean seeds in water swelled as they absorbed water and broke through the plastic wrap.)

How do you think that plants and plant roots affect the soil? (They cause physical weathering by growing into hard soil and the cracks of rocks and wedging them apart.)

Name one way that plants help the soil. (Leaves and plant parts decay, adding humus and nutrients to the soil.)

Name one way that soil helps humans. (Soil makes it possible for plants to grow, and we eat these plants; we also eat animals that depend upon plants for food.)

VIDEO NOTES
Soils

Directions Match the definitions on the left with the terms on the right.

_____ 1. Mixture of weathered rock and organic material on Earth's surface

_____ 2. Made from the organic matter of plants and animals

_____ 3. Mixture of clay, sand and humus

_____ 4. Compact, water-holding soil

_____ 5. Loose, grainy soil

a. clay
b. sand
c. soil
d. loam
e. humus

Circle T if the statement is true. Circle F is the statement is false.

T *or* **F**   6. Living plants and animals help soil by pushing through the ground and breaking up hard soils.

T *or* **F**   7. When plants and animals die and decay they become an important part of the soil called humus.

T *or* **F**   8. When soil forms, it forms in an even layer on Earth's surface.

T *or* **F**   9. Flat lands have deeper layers of soil while hills and mountains have thinner layers of soil.

T *or* **F**   10. Soil is the same everywhere in the world.

List five important things that we would not have without soil:

1. _____

2. _____

3. _____

4. _____

5. _____

ACTIVITY

Soils

Find Out How do animals and plants use soil to survive?

What to Do Picture yourself as an animal such as an earthworm, mole, prairie dog or groundhog living in the soil and answer the following questions.

1. What do you need from the soil in order to live? _____

_____

_____

_____

_____

2. How does the moisture, dryness, hardness or softness of the soil affect you?_____

_____

_____

_____

_____

3. What do you eat? How does the soil provide your food?_____

_____

_____

_____

_____

4. What does the soil provide to both animals and plants?_____

_____

_____

_____

_____

Earth Beneath You
Natural Resources

The Big QUESTION *Why are the Earth's natural resources important?*

Media Resources
Videotape
Grade 3 • Earth Science

Laserdisc
Grade 3 • Earth • Side 1

SAMPLE

Frame 12345

Objectives
- name different kinds of natural resources
- explain why natural resources need to be protected

Vocabulary
resource, renewable, nonrenewable, inexhaustible

1 Introduce

Lesson Background A natural resource is any naturally-occurring object or material that living things find useful or necessary. In this lesson, students will identify and describe the importance of Earth materials and classify them as renewable, nonrenewable or inexhaustible natural resources. Among the most dangerous problems facing our world is the destruction of the environment because of misuses of resources. Renewable resources (water, air, fish, trees) are being used at such a rapid rate that they are not able to replace themselves fast enough. Nonrenewable resources, such as soil and some rocks and minerals (coal, petroleum), are being depleted and used in ways that pollute our environment. On the other hand, inexhaustible resources such as sunlight, wind and moving waters are not always used as much as they should be. Inexhaustible resource utilization (windmills, solar power, hydroelectric power) can be very valuable alternatives to more wasteful energy producers (machines that use fossil fuels).

Activate Prior Knowledge Ask students to name things that they use every day. As each item is named, write it on the chalkboard.

2 Teach

Before Viewing the Video
For Discussion
Tell students to imagine a library where they could only check out a book once. Since the books are nonrenewable, how would this affect their reading habits? Referring to the list on the chalkboard, ask students whether each item can be used again. Tell students to listen for a definition of "natural resources" in the video and to be prepared to define the term in their own words.

After Viewing the Video
Thinking Critically
Next to each item written on the chalkboard, write down students' responses as to whether that resource is renewable, nonrenewable or inexhaustible. How could they shift to using more inexhaustible resources?

Using Video Notes
Distribute copies of Page 72 to students. Show the video again and have students fill in the lists on their worksheet.

③ Close

Checkpoint
- What are the three different kinds of natural resources? (renewable, nonrenewable, inexhaustible)
- Why should natural resources be protected? (Since some natural resources are nonrenewable and others are used too fast to be renewed in time, they should be protected for future use.)

Activity
What to Expect Students will identify ways to change some of their regular activities to conserve natural resources.

Distribute copies of Page 73 and tell students to suggest two or three environmentally friendly alternatives to each action that uses natural resources.

Alternate Activity
Find Out How is recycled paper made?

What You Need scrap paper, graduated cylinder, a large basin, egg beater, hot water, laundry starch, 30-mesh fine screen, rolling pin, blotting paper

Teacher Preparation Tear scrap paper into small pieces. Fill a graduated cylinder to the 125-mL mark with shredded paper. Place the shredded paper in the basin and add 500 mL hot water and 30 mL of starch. Make a batch large enough for your class. Beat the mixture with an egg beater until the pulp is the consistency of a light gravy.

What to Do Have students dip the screen into the pulp until the screen is completely coated. Then, help them place the pulp-coated screen between two sheets of blotting paper and press out the excess water with a rolling pin. Turn it all over and remove the top blotter and screen. Allow the paper to dry on the bottom blotter. Help students carefully peel the recycled paper from the remaining blotter when dry.

Conclusions
What natural resource is used to make paper? (trees) What paper products do you use often? (school paper, paper cups and plates, food packaging, facial tissue, paper towels, toilet paper) We try to reuse paper whenever possible so we won't have to cut down too many trees. What are some ways you can reuse paper? (Accept all reasonable answers.)

FUN FACT
The average American uses eight times as much fuel energy as an average person anywhere else in the world.

ACROSS THE CURRICULUM
Health
Salt is one of the most abundant minerals on Earth. It's in our food, our ocean water, and our blood, sweat and tears. You could not live without salt. It controls the amount of water in your cells so they don't burst. It also helps conduct your body's electrical signals that make your muscles contract and your heart beat. However, too much salt can be bad for your health.

Have students look at the labels on food products at home, and look for the word *sodium*—one of the elements that makes up salt. Have them make a chart of foods that contain sodium and record the amount of sodium for each kind of food. How many products have salt in them? Are they surprised?

VIDEO NOTES

Natural Resources

Directions In your own words, define "natural resources": _____

_____

_____

_____

Match each resource to the correct type.

Renewable	trees
	minerals
	sunlight
	water
	soil
	wind
Nonrenewable	air
	rocks
	moving water
	fish
Inexhaustible	coal
	animals
	petroleum

ACTIVITY
Natural Resources

Find Out What can I do to conserve natural resources?

What to Do

Circle the type of resource that is used in each action. Describe alternative actions you could do to conserve the natural resources.

Petroleum

Renewable

Nonrenewable

_____

_____

_____

_____

Trees

Renewable

Nonrenewable

_____

_____

_____

_____

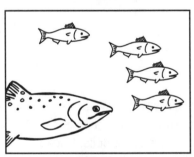

Fish

Renewable

Nonrenewable

_____

_____

_____

_____

Water

Renewable

Nonrenewable

_____

_____

_____

_____

How Matter Changes
Properties of Matter

The Big QUESTION *How can you describe matter?*

Media Resources
Videotape
Grade 3 • Physical Science

Laserdisc
Grade 3 • Physical • Side 1
SAMPLE
Frame 12345

Objectives
- identify the building blocks of matter and elements
- describe the three states of matter
- explain how states of matter can change

Vocabulary
atom, element, oxygen, property, evaporation

① Introduce

Lesson Background Atoms are tiny pieces of matter that are characterized as the "building blocks" of matter. Elements are simple substances that are constructed of just one type of atom. Matter is generally described as existing in one of three states: solid, liquid or gas. Atoms are packed in a tight, orderly group in solids, where they do not move very much. In liquids, atoms are grouped more loosely and move randomly. In gases, the atoms are far apart from each other and thus move easily and quickly. Furthermore, solids hold their shapes while liquids and gases conform to the shapes of their containers. Liquids can be held in an open container whereas gases must be completely enclosed. Types of matter can often be differentiated by physical properties. Physical properties include color, size, shape, texture, hardness, melting point, odor and the ability to dissolve in water. Physical properties are used to identify observable types of matter; chemical properties are required to differentiate between specific types of matter such as the colorless, tasteless, odorless gases known as hydrogen and oxygen.

Activate Prior Knowledge You may want to begin this unit by asking students what they think matter is. Write their responses on the chalkboard and refer back to this list or change information as concepts are learned.

② Teach

Before Viewing the Video
For Discussion
Ask students questions about where they are sitting. What is between you and the ceiling? (air) What is your desk made of? (wood, metal, glue, etc.) What is in your desk? (pens, pencils, books, paper, etc.) List their responses on the chalkboard. Ask what the items listed all have in common. (They're all made of matter.) Show the video and tell students to watch and listen carefully to learn more about matter.

After Viewing the Video
Thinking Critically
1. Move to a clear part of the chalkboard and write columns as students list the three states of matter (solid, liquid, gas). Encourage students to match each item in the earlier list to an appropriate column (glue would fit under liquid, wood under solid, air under gas, etc.).

2. Choose a few items from the list and ask students to name its physical properties. Since air does not have obvious observable properties, ask students how they could tell the difference between a balloon filled with oxygen versus a balloon filled with helium (since helium is lighter than air, the balloon filled with helium would float; weight, or density, is a physical property).

3. Finally, ask students to once again name what every item on the chalkboard has in common (all items are made up of matter and all matter is made up of atoms). This might be a good opportunity to address two common and related misconceptions, that is, that matter can be seen and that everything we can see is matter. Actually, air cannot be seen but is made of matter, and light can be seen but is not made of matter.

Using Video Notes
Distribute copies of Page 76 to students. Give students several minutes to review the statements. Then, show the video once again. Have students fill in the blanks on their worksheet.

❸ Close
Checkpoint
- What are the building blocks of matter and elements? (atoms; elements are made of only one kind of atom)
- What are the three states of matter? (solid, liquid, gas)
- How can water change state? What happens to the atoms? (A liquid can change into a solid by freezing and into a gas (water vapor) by boiling; atoms in solid objects are packed tightly together in an orderly pattern; atoms in a liquid move in a jumbled, disorderly way; atoms in a gas are very far apart and move easily and quickly.)

Activity
What to Expect Students will demonstrate that matter takes up space.

Distribute copies of Page 77. Arrange students into groups of four. Provide them with activity materials and help them follow the instructions. Students should record their findings as they conduct the experiment.

VIDEO NOTES

Properties of Matter

Directions Write the letter of the answer that completes each sentence.

1. One gas found in the air we breathe is ___ .
   - a. oxygen
   - b. water
   - c. energy
   - d. iron

2. The tiniest piece of any kind of matter is a(n) ___ .
   - a. liquid
   - b. gas
   - c. atom
   - d. property

3. A characteristic, or something that makes one thing different from another, is a(n) ___ .
   - a. state
   - b. property
   - c. atom
   - d. element

4. Solids, liquids and gases are the three ___ of matter.
   - a. colors
   - b. states
   - c. sizes
   - d. properties

5. All ___ hold their own shape and can be described by their color, texture and size.
   - a. liquids
   - b. solids
   - c. gases
   - d. elements

6. ___ do not hold their own shape, but they can be described by their color and texture.
   - a. liquids
   - b. solids
   - c. gases
   - d. elements

7. ___ do not hold their own shape, and their atoms move easily and quickly.
   - a. liquids
   - b. solids
   - c. gases
   - d. elements

8. ___ is an element because it is made up of only one kind of atom.
   - a. matter
   - b. gas
   - c. silver
   - d. carbonated drink bubbles

ACTIVITY

Properties of Matter

Find Out Does matter take up space?

What You Need 100-mL graduated cylinder (one per group), marbles (15 per group), water

What to Do

1. Pour 40 milliliters (mL) of water into the graduated cylinder. Record the water level in the chart below. Then, drop one marble in the cylinder and record the water level in the chart. Add four more marbles, then record the water level in the chart.

2. Predict what the water level will be with 10 marbles and 15 marbles.

3. Test each of your predictions with the marbles and the cylinder and complete the chart.

Data Chart	
 Number of marbles	Water level (in mL)
0	
1	
5	
10	
15	

Talk About It How much air is left in the graduated cylinder?

Lesson 1 *Properties of Matter* **77**

How Matter Changes
Combining Substances

The Big QUESTION *What happens when you combine different substances?*

Media Resources
Videotape
Grade 3 • Physical Science

Laserdisc
Grade 3 • Physical • Side 1
SAMPLE
Frame 12345

❶ Introduce

Lesson Background Most matter does not consist of pure elements. The elements are usually combined into more complex substances that include mixtures and compounds. Mixtures result when substances mix loosely and are not chemically or physically altered (oil and water, iron filings and sand). Solutions are mixtures in which one substance dissolves in another (salt water), therefore altering a substance's physical properties. The substances in mixtures and solutions can be separated through evaporation, filtration or other physical means. Compounds result when two or more substances combine chemically to form a new substance (tarnish, rust, sour milk, ashes from burned wood). Unlike the physical changes involved in mixtures and solutions, chemical changes are irreversible (the materials cannot be separated back to the original substances).

Activate Prior Knowledge Ask students to name food items that contain substances that are mixed together *(fruit and cereal and milk, cookie dough, lemonade)*. Write these items on the chalkboard.

❷ Teach

Before Viewing the Video
For Discussion
On a clear part of the chalkboard, and leaving room for one more row at the top, draw three columns headed by "Mixture," "Solution" and "Compound." Choose one food item that serves as an example for a mixture and explain that substances in mixtures stay the same. Challenge students to name the other food items on the chalkboard that they think are mixtures. Choose a food item that serves as an example for a solution and explain that one substance dissolves into another to form a solution. Challenge students to name the other food items on the chalkboard that they think are solutions. Now, tell students to watch the video to find out what happens to substances that combine to form a compound.

After Viewing the Video
Thinking Critically
1. Add two more columns above the three existing columns: write "Physical" above "Mixtures" and "Solutions," and write "Chemical" above "Compounds." Explain that a physical change just affects the physical properties of a substance — it may change shape or color, but it is still the same substance.

Objectives
- distinguish between elements and compounds
- describe what happens when substances combine
- distinguish between physical and chemical changes

Vocabulary
compound, substances, mixture, solution

On the other hand, a chemical change turns two or more substances into a completely new substance—two gases (hydrogen and oxygen) can combine to make a liquid called water.

2. Ask students to assign the appropriate food items on the chalkboard to the "Compounds" column and prompt them to explain how each substance in each combination changed.

Using Video Notes

Distribute copies of Page 80 to students. Have them watch the video once again. This time, have students watch for clues to the questions on their worksheet.

3 Close

Checkpoint

- What is the difference between elements and compounds? (Elements are made of only one kind of atom whereas compounds are chemical combinations of two or more kinds of atoms.)

- What happens when substances combine? (Substances combine to form one of three things: (1) a mixture, in which neither substance is altered, (2) a solution, in which one substance dissolves into another, or (3) a compound, in which two or more substances combine to form a new substance.)

- What is the difference between physical and chemical changes? (Substances in physical changes retain their identity and can be separated back to their original form; substances in a chemical change become a new substance and cannot be returned to their original form.)

Activity

What to Expect Students will illustrate the chemical combinations that form water and salt, and they will describe how the substances in each compound change.

Distribute copies of Page 81. Discuss chemicals and formulas with students, specifying that salt is sodium chloride, a chemical combination of one atom of the element sodium with one atom of the element chlorine. Water is a compound composed of two atoms of the element hydrogen and one atom of the element oxygen.

FUN FACT

Some compounds break down very easily. In fact, hydrogen peroxide ($2H_2O_2$) turns into hydrogen and oxygen so readily that peroxide bottles must be dated. On the other hand, water (H_2O) can only be broken down into hydrogen and oxygen by using electricity.

ACROSS THE CURRICULUM

Social Studies

Salt, which has many uses, is mined all over the world and is even obtained from seawater by evaporation. The saltiest body of water in the world, the Dead Sea, is nearly 25 percent salt. All this salt makes the water so dense that it is almost impossible to remain submerged in it. Have students research the Dead Sea and write about what kinds of animals live in it.

VIDEO NOTES

Combining Substances

Directions Match each term to its definition.

1. A combination of two or more substances that can be separated very easily: _____

2. A combination of one substance dissolved into another substance: _____

3. A combination of two or more substances that results in a new substance: _____

substances

mixture

solution

compound

4. Basic materials that make up all things: _____

Directions Match each substance to the correct kind of combination.

rust •

chocolate milk •

water •

mixture

cereal and milk •

oil and water •

solution

salt water •

tarnish on a penny •

wood ashes •

compound

table salt •

ACTIVITY
Combining Substances

Find Out What elements are salt and water made of?

What You Need crayons, scissors, glue, construction paper

What to Do

Color each type of atom a different color. Cut out the atoms and glue them together on a piece of construction paper to form water (two hydrogen atoms plus one oxygen atom) and salt (one sodium atom plus one chlorine atom). Answer the questions below.

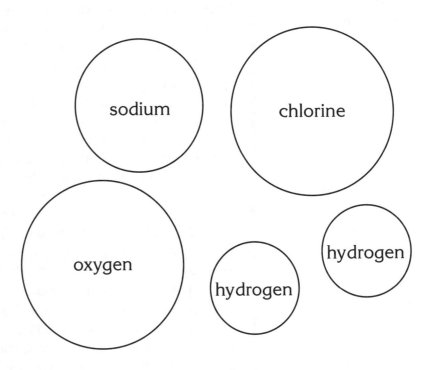

What are the names of the elements you used? _____

_____

What are the names of the compounds you formed? _____

_____

Both hydrogen and oxygen are invisible gases. How do they change when they are chemically combined to make water? _____

_____

Sodium is a soft white metal and chlorine is a poisonous, greenish gas. How do they change when they are chemically combined to make salt?

_____

_____

How Matter Changes
Energy

The Big QUESTION *What happens to energy during a chemical reaction?*

Media Resources
Videotape
Grade 3 ● Physical Science

Laserdisc
Grade 3 ● Physical ● Side 1
SAMPLE
Frame 12345

1 Introduce

Lesson Background Energy is the ability to do work. All life depends upon energy, and most of the energy on Earth comes from the sun. The sun enables plants, which serve as fuel for humans and animals to grow, and the sun's energy is stored up in types of fuel, such as coal, wood and oil. The sun also evaporates water that falls again as rain and forms rivers and lakes that can be used for energy. Additionally, humans harness the sun's rays in various solar-energy applications. In this lesson, students are introduced to stored energy (potential energy) and the energy of movement (kinetic energy). Stored energy represents work that is already done. For example, a rock on the edge of a cliff has stored energy, and so does a battery. Energy is constantly transformed from one form to another to do work. For example, the stored energy of gasoline is burned in an automobile engine and converted into (kinetic) mechanical energy that powers a car.

Activate Prior Knowledge Ask students about stored energy. Does a moving ball have energy? Why? (Yes, because it has the ability to make something move.) How could you show that a ball has energy? (Possible answers include rolling a ball into an object, such as a toy soldier, and knocking it down.) When the ball is at rest on a table, does it still have energy? Why? (Yes, it has potential energy due to gravity.)

Objectives
- define stored energy
- describe how energy changes form
- identify ways in which people use energy

Vocabulary
energy, heat, fuels

2 Teach
Before Viewing the Video
For Discussion
Ask students to name other objects that have energy. How can you tell that each object has energy? Where does the energy come from? Write these responses on the chalkboard. Tell students to watch and listen carefully for information about energy and its various forms.

After Viewing the Video
Thinking Critically
1. Review the list on the chalkboard and ask students to explain whether each object has stored energy (called potential energy) or energy of motion (called kinetic energy). Make sure that students recognize that an object can have both kinds of energy.

2. Add any additional forms of energy that students recall from the video.

Using Video Notes

Distribute copies of Page 84 to students. Have them watch the video once again. This time, have students complete the chart on their worksheets.

❸ Close

Checkpoint

- What is stored energy? (potential energy that is stored in an object and not yet used)

- What happens to the form of energy used in turning on a flashlight? (Chemical energy stored in the batteries turns into electrical energy which becomes heat and light.)

- How do people use energy? (The human body uses food as fuel to allow it to live, move and grow. Also, people use energy for all kinds of everyday activities such as driving a car, using a computer, etc.)

Activity

What to Expect Students will predict and experiment to observe the use of stored energy in a balloon.

Distribute Page 85 and one balloon to each student. Tell students to answer the questions on their Activity Page by using balloons and the scientific method.

Alternate Activity

Find Out What kind of energy puffs up biscuits?

What You Need plastic spoon, 1 level spoonful of baking soda, goggles, 4 spoonfuls of vinegar, paper funnel, 1-L bottle, balloon

What to Do

Put on your goggles. Put the vinegar in the 1-L bottle. Make a paper funnel and use it to put the baking soda inside the balloon. Being careful not to drop any baking soda in the bottle, put the balloon neck around the bottle neck. Now, lift the balloon up and shake the baking soda down into the vinegar.

Note: You could also allow students to conduct this experiment in groups of two. Make sure that they wear safety goggles and aprons, and warn them that vinegar and baking soda are caustic substances that could sting and burn in cuts and scrapes.

Conclusion

Was energy released? How can you tell? (Yes, the energy released caused the balloons to fill up.)

FUN FACT

A calorie is the amount of energy that is needed to raise the temperature of one gram of water by one degree Celsius.

ACROSS THE CURRICULUM

Art

Have students make a collage of home products that use energy and another collage of those that do not. Which nonconsuming-energy products perform the same activity as energy-consuming products?

Language Arts

Have students find information on the 1979 oil shortage in the United States. Ask students to write an essay on how a new oil shortage in the United States would affect their lives.

VIDEO NOTES

Energy

Directions As you watch the video, observe and record the energy transformations involved in the work being accomplished.

	Energy In	Work	Energy Out
Car	Gasoline	Driving	Mechanical
Wind-up toy			
Candle			
Match			
Campfire			
Flashlight			
Electric stove			
Human body			

ACTIVITY

Energy

Find Out How can you store energy in a balloon?

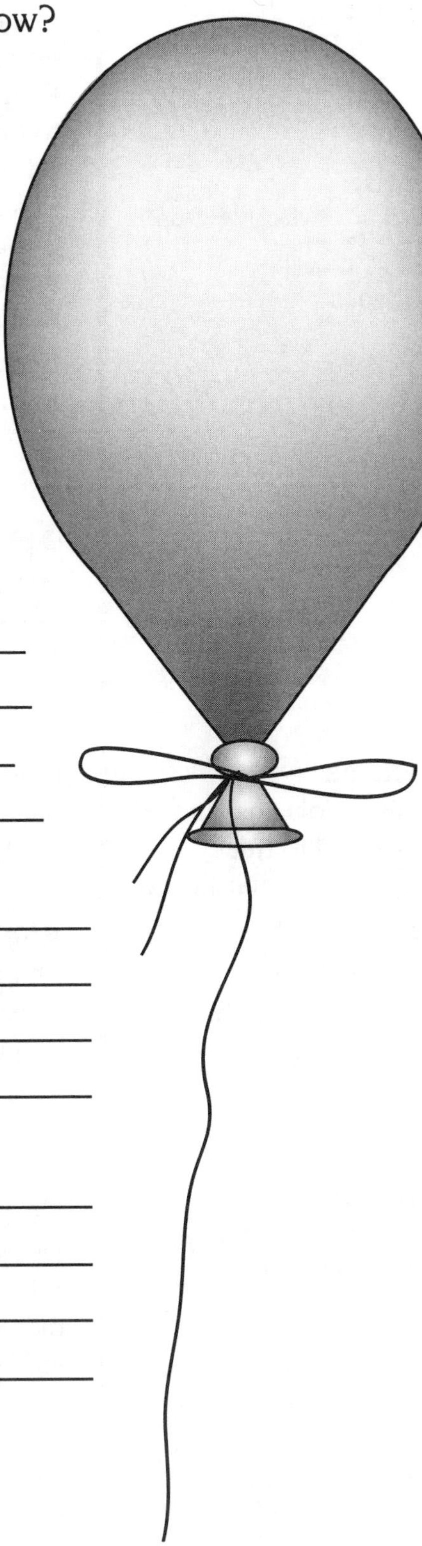

1. Does your balloon have energy? How do you know?

_____

_____

_____

_____

2. Predict what will give the balloon more energy.

_____

_____

_____

_____

3. How did you test your prediction?

_____

_____

_____

_____

4. What did you observe?

_____

_____

_____

_____

5. How did you give your balloon stored energy?

_____

_____

_____

_____

Light
Light Creates Changes

The Big QUESTION *How does light make changes?*

Media Resources
Videotape
Grade 3 ● Physical Science

Laserdisc
Grade 3 ● Physical ● Side 1
||||| SAMPLE |||||
Frame 12345

1 Introduce

Lesson Background Light is a form of energy, just like sound and heat. Most of the energy on Earth is derived from the sun. People use light for energy in many forms, including solar energy to heat homes. Light normally travels in straight lines and reflects off the objects it hits. This reflected light enables the human eye to see. The pupil regulates the amount of light allowed to pass through the cornea and the lens to the retina, where an upside-down "picture" is produced. The brain then corrects the image so that we see objects as they exist.

Activate Prior Knowledge Assess students' prior knowledge by asking them the following questions: What is light? How does light travel? How does light help us see?

2 Teach

Before Viewing the Video
For Discussion

Ask students to name different ways that people use light. Write their responses on the chalkboard and ask them to think of how these sources produce light.

Explain to students that we see objects because light reflects off objects and into our eyes. Tell students to watch the video to find out how eyes work.

After Viewing the Video
Thinking Critically

1. Ask students to name the different parts of the human eye.
   (pupil, cornea, lens, retina)

2. Ask students to think of different ways that solar energy could be used.

Using Video Notes
Distribute copies of Page 88 to students. Give them several minutes to review the words and statements. Have them fill in any blanks they can. Then, show the video again. As they watch, have students complete their worksheets.

Objectives
- explain how light is converted to heat energy in a solar home
- state that light travels in a straight line
- describe how light allows us to see

Vocabulary
solar energy, cornea, pupil, lens, retina

❸ Close

Checkpoint

- How is light converted to heat energy in a solar house? (Glass panels on the roof contain water pipes that are warmed by the sun's rays. This hot water circulates through other pipes that warm the entire home.)

- How does light travel? (Normally, light travels in straight lines.)

- How does light help us see? (Our eyes convert light that bounces off objects into an image our brain can read.)

Activity

What to Expect Students will identify the path light travels through the human eye to produce a visual image.

Distribute Page 89 to students and tell them to label each part of the diagram. Make sure that they draw an upside-down image in the retina.

Alternate Activity

Find Out How are heat energy and light energy related?

What You Need desk lamp, 15-cm circle of aluminum foil cut into a spiral, needle and 15-cm piece of thread

What to Do

1. Pull the needle and thread through the center of the aluminum foil spiral. Knot the thread.

2. Hold the thread up so that the aluminum spiral falls a few inches above the bulb of the desk lamp. Turn on the lamp.

3. Have students observe what happens. (The aluminum spiral twirls as the heated air moves upward.)

4. Move the spiral away from the lamp. Have students observe what happens. (The spiral slows down and stops.)

Conclusions

What is your observation? (Heated air moves upward.)

What does this show about the relationship of heat and light? (Light is often accompanied by heat. In this example, heat energy produces light energy.)

FUN FACT

By the time light from the sun gets to Earth, it is eight minutes old. Just imagine how far away a light-year is . . . and most stars are millions of light-years away!

ACROSS THE CURRICULUM

Health

How can energy from the sun be harmful? Have students use the Internet or media center to research heat stroke, sunburn and skin cancer.

VIDEO NOTES
Light Creates Changes

Directions Fill in the blanks with the correct word.

1. The energy we get from the warmth of the sun is
   _____ .

2. A straight line of light is called a(n)
   _____ .

3. Shadows are formed when a(n) _____
   object blocks light.

4. We are able to see objects because they
   _____ light.

5. The _____ regulates how much light
   enters the human eye.

6. The _____ contains cells that help us
   see in color.

7. Earth gets the majority of its light and energy
   from _____ .

8. _____ light comes in many forms, such as electric light and
   neon light.

9. Light is a form of _____ .

opaque

ray

pupil

the sun

solar energy

energy

reflect

retina

artificial

ACTIVITY

Light Creates Changes

Find Out What are the major parts of the human eye?

What to Do Label the parts of the eye that light travels through to create a visual image.

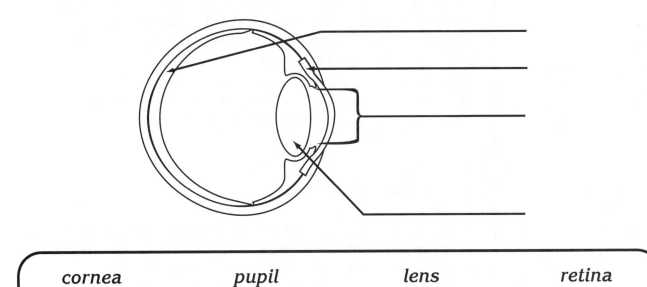

| cornea | pupil | lens | retina |

Draw the image as it appears on the retina before the brain corrects it.

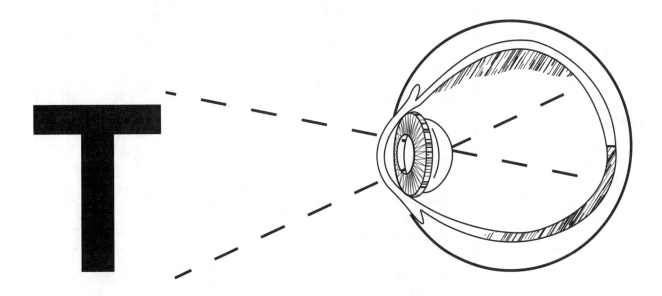

Light
Properties of Light

The Big QUESTION *How does light behave?*

Media Resources
Videotape
Grade 3 • Physical Science

Laserdisc
Grade 3 • Physical • Side 1
SAMPLE
Frame 12345

Objectives
- describe how light travels
- explain how shadows are made
- describe how light is reflected

Vocabulary
transparent, translucent, mirror, scatter

1 Introduce

Lesson Background Materials that allow most light to pass through, such as clear glass, are described as transparent. Waxed paper and frosted glass are examples of materials that allow only some light to pass through; those materials are translucent. Materials like brown paper and many solid objects that allow no light to pass through are opaque.

Activate Prior Knowledge Ask students to describe what happens to light when it encounters certain objects—mirrors, clear glass, sunglasses, dust in the air, muddy water, etc.

2 Teach

Before Viewing the Video
For Discussion
Ask students to look around the classroom for objects that reflect light easily. Have students name the objects they find and explain what all the reflecting objects have in common. How are reflecting objects useful? Do they know what a reflection really is? Have students watch and listen carfully to learn more about reflections and other properties of light.

After Viewing the Video
Thinking Critically
Have students define "opaque," "transparent" and "translucent." Ask them to list objects in and out of the classroom that fit these descriptions. Write their responses on the chalkboard under appropriately labeled columns.

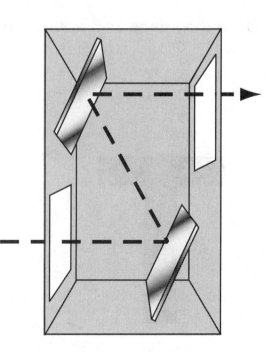

Using Video Notes
Distribute copies of Page 92 to students. Have them watch the video once again and listen for clues to answer the questions on their worksheet.

3 Close

Checkpoint

- What are the different kinds of telescopes? (Optical telescopes, either refracting or reflecting, use lenses or mirrors to collect light and produce an image. Radio telescopes detect radio waves transmitted by objects in space.)

- What is starlight? (The emission of electromagnetic energy that is both visible and invisible. The visible light is a mixture of colors that travel through space at the speed of light.)

- What are the phases of a star's life? (A star may form from a nebula or from clouds of hydrogen dust and gas. Over time, the matter collapses due to gravitational forces until the temperature and mass increase to the point that a star begins to shine. Eventually, the hydrogen will burn off and the star dies by either cooling down or exploding into a supernova.)

- What is a galaxy? (It is a group of billions or trillions of stars. They can be either spiral, irregular or elliptical in shape.)

Activity

What to Expect Students will demonstrate their knowledge of the life cycle of a star and the different characteristics associated with each life stage.

Distribute copies of Page 93. Have students complete the two parts of the activity. In the first section, students will fill in the blanks using the word bank. Then, students will apply what they learned from the video to answer the questions.

FUN FACT

Light from the closest star beyond the sun takes more than four years to reach Earth.

ACROSS THE CURRICULUM

Art—Telescoping Vision
To demonstrate how the aid of instruments has allowed people to examine planets, stars and other celestial objects in the sky, have students draw three pictures. One picture will show a distant view of space from Earth. The next will show a small part of the first picture viewed from a telescope. The final picture will show a photographic view taken from the Hubble space telescope of an area within the second picture. Each picture should have a closer view and more detail. Allow students to display their pictures.

VIDEO NOTES

Properties of Light

Directions Answer the following questions.

1. Why can you see yourself in the mirror?_____

_____

_____

2. Why do automobiles, bicycles and other transportation vehicles have mirrors?_____

_____

_____

3. What do you think is the difference between natural and artificial light?

_____

_____

_____

_____

4. Why can you see through a windshield but not through a car seat?

_____

_____

_____

5. Is a frosted light bulb transparent, translucent or opaque? How do you know?

_____

_____

_____

_____

_____

ACTIVITY
Properties of Light

Find Out What happens to light when it hits the water?

What You Need coin, water, opaque cup, tape

What to Do

Tape a coin to the inside bottom of the opaque cup. Stand back until the coin is just hidden from your sight. Now have your partner slowly pour water into the cup. What happens? _____

_____

In the space below, draw your picture of the cup with the coin in it and yourself looking at the coin. Does light bend? _____

Talk About It Light waves normally travel in straight lines. Why can you see the coin after the water is poured in the cup?

Light
The Spectrum of Light

The Big QUESTION *What are the named colors in the spectrum of light?*

Media Resources
Videotape
Grade 3 ● Physical Science

Laserdisc
Grade 3 ● Physical ● Side 1
SAMPLE
Frame 12345

1 Introduce

Lesson Background Visible light is one form of electromagnetic energy. The frequency of a light wave determines its color, — red has the lowest frequency and violet has the highest frequency. There are three primary colors of light—green, blue and red. Mixing these colors in various ways will produce any color of light, and mixing all three together produces white light. Mixing all three primary pigments (yellow, blue and red) produces black. Color is perceived by subtraction. In other words, the color of an object is determined by the wavelength of visible light that is not absorbed — green objects absorb every color except green, which is reflected. Black and white are often erroneously thought of as noncolors. In actuality, white objects are reflecting all colors of light to the eye while black objects absorb all colors of light, reflecting very little light to the eye.

Activate Prior Knowledge Ask students to name all the colors in the rainbow. Write their responses on the chalkboard. Where do these colors come from?

2 Teach

Before Viewing the Video
For Discussion
Ask students to think about black and white. Do these count as colors? Have students watch and listen for more information about light and colors.

After Viewing the Video
Thinking Critically
1. Referencing the list on the chalkboard, ask students to help you revise the list according the "Roy G Biv" memory device covered in the video.
2. Now ask students to list the colors that are included in the color white and the color black. How are black and white different? How are they the same?

Using Video Notes
Distribute copies of Page 96 to students. Have them watch the video once again. Students should first color the full spectrum as it strikes the snowman, tire and lips on their worksheet. Then, students should show which wavelengths are reflected back to the eye.

Objectives
● explain how a prism separates light
● describe a spectrum
● explain how we see colors

Vocabulary
prism, refract, spectrum

3 Close

Checkpoint

- What does a prism do? (A prism is a clear object, often triangular, with flat sides, that catches the sun's light and breaks it into different colors.)

- What does the visible light spectrum look like? (The spectrum is a band of seven main colors: red, orange, yellow, green, blue, indigo, violet.)

- How do we see colors? (Objects absorb every color of the spectrum except the color that we see – a red apple absorbs orange, yellow, green, blue, indigo and violet light waves and reflects red.)

Activity

What to Expect Students should infer that colors mix together to form white light.

Distribute Page 97 to students and review the concept that white light is a combination of all colors. Make sure that students spin the wheel very quickly to observe the colors mixing.

Alternate Activity

Find Out How is a rainbow formed?

What You Need sunshine, flat mirror, shallow dish full of water, white card, tape

What to Do Help students simulate a rainbow by following these directions:

1. Rest the flat mirror against the inside of the dish.

2. Position the dish so that sunlight falls onto the mirror.

3. Hold the white card in front of the mirror and move it around until a rainbow of colors appears. You may have to adjust the position of the mirror to get it just right.

Conclusion

How is the sunlight split in this experiment? (The sunlight is split because the wedge of water between the surface of the water and the mirror acts as a prism. Prisms work by bending, or refracting, light rays.)

FUN FACT

A rainbow can be seen only in the morning or late afternoon and it can occur only when the sun is 40 degrees or less above the horizon.

ACROSS THE CURRICULUM

Art

Artists use light to create different effects. Encourage students to study a famous artist or style of art, such as Impressionism, to discover how artists used light in their paintings. Perhaps the art teacher could help with this activity.

VIDEO NOTES

The Spectrum of Light

Directions Use crayons, colored pencils or markers (red, orange, yellow, green, blue and violet) to fill in the colors that each object absorbs. Then fill in the colors that each object reflects.

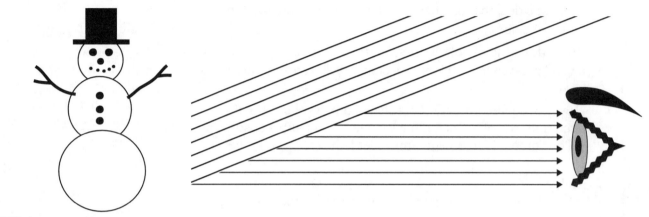

White snowman

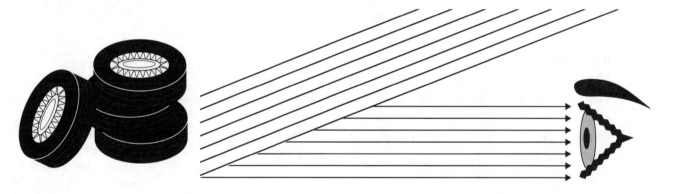

Black tire

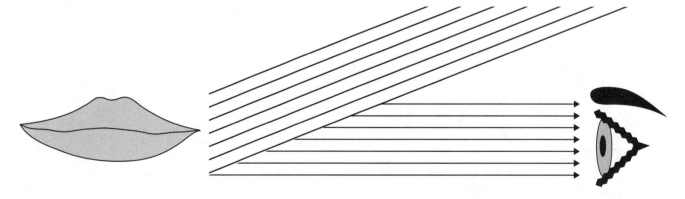

Red lips

ACTIVITY
The Spectrum of Light

Find Out What is white light made of?

What You Need heavy cardboard disk 10 cm in diameter, thick rubber band, colored pencils (red, orange, yellow, green, blue and violet), scissors, glue

What to Do

Color the sections in the disk below as indicated. Cut out the circle and glue it to the cardboard disk. Now make two small holes in the middle of the disk, about 1 cm apart. Thread the rubber band through the holes and make a knot on each side. Twist the rubber band and pull it tight. What do you see as the disk spins?

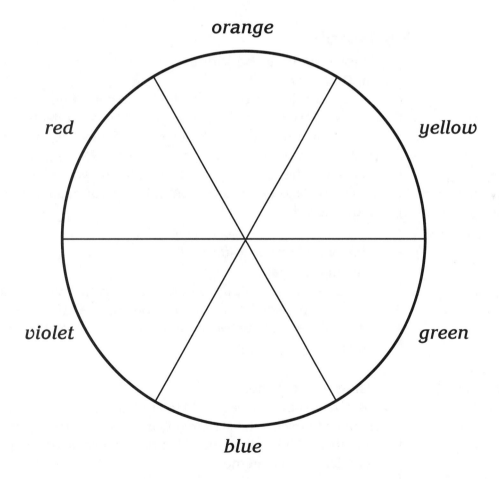

Talk About It What happens when colored objects move very fast? Could you still see all the separate colors? What color did you see?

Simple Machines & How They Work
Force and Work

The Big QUESTION *How are force and work related?*

Media Resources

Videotape
Grade 3 • Physical Science

Laserdisc
Grade 3 • Physical • Side 1
SAMPLE
Frame 12345

Objectives
- define force
- explain how work is done
- describe how to measure force

Vocabulary
force, gravity, speed, work, distance

---

1 Introduce

Lesson Background All pushes and pulls are forces. When work is done on an object, a continuously applied force moves the object over a distance. In this lesson, students will learn that to do any work at all, some amount of force is needed.

Activate Prior Knowledge Have students think about machines they use every day. Record their ideas on the chalkboard. Ask students to summarize what their various machines do. Write the verbs they express beside the name of each machine. Then, review the list of verbs. Are there any common themes? Could these verbs be consolidated under umbrella verbs such as "push" and "pull"? Try it!

2 Teach

Before Viewing the Video
For Discussion
Make two columns beside the list of machines on the chalkboard. Label one "push" and the other "pull." Have students analyze each of their machines and determine if they are, in essence, pushing something or pulling something. Do the machines themselves need to be pushed or pulled to work? Tell students to watch and listen for more information about pushes and pulls. Show the video.

After Viewing the Video
Thinking Critically
Have students recall things from the video that were either pushed or pulled. Add these to the list on the chalkboard. What are the differences between performing a push versus performing a pull?

Using Video Notes
Distribute copies of Page 100 to students. Allow students several minutes to review the statements and vocabulary words on their worksheet. Then, show the video once again so students can complete each statement.

3 Close

Checkpoint

- What is force? (A force is a push or a pull.)
- How is work done? (Work is done when a force makes an object move over a distance.)
- How can force be measured? (Weight is how we measure the force of gravity; speed is how we measure how fast an object is moving; speed is computed by dividing distance by time.)

Activity

What to Expect Students will identify forces acting on simple machines as either pushes or pulls.

Distribute Page 101 to students. Have them analyze the various illustrations and tell whether a "push" or "pull" is being applied. Then, have students briefly explain the work that each simple machine does.

Alternate Activity

Find Out How much force is required to do work?

What You Need a small wagon, spring scales calibrated in newtons

What to Do

Bring a small wagon to class. Have students experiment with the wagon to discover when work is being done and the amount of force required to do the work. Tell students to predict which of the following activities requires more force, then test their predictions either qualitatively, by asking questions such as *How much work do you think was done? A lot? Some? A little?* or quantitatively, by using spring scales and measuring force in newtons.

1. Pull one student up a hill or pull two students of similar size on flat ground.
2. Pull one larger student or pull two smaller students.
3. Pull one tall, thin student or pull one short, muscular student.

Have students repeat the above activities, but, this time, have them push the wagon rather than pull it.

Conclusion

Which action requires less effort, pushing or pulling the wagon? (pushing)

Why does extra weight in the wagon make it harder to move? (More weight increases the load and increases the amount of force required to move the wagon.)

VIDEO NOTES

Force and Work

Directions Watch the video. Listen for clues to help you fill in the blanks. Choose words from the list below.

1. _____ is the measure of how fast an object moves over a distance.

2. _____ is what happens when something changes position.

3. _____ is done when a _____ makes an object move over a distance.

4. _____ is how far something moves.

5. If something is moving, its position is_____ .

6. No work is done on an object if the object does not _____ .

7. A simple _____ is one with few or no moving parts.

8. _____ is the effort made when matter is pushed or pulled.

9. _____ is used or is changed to exert force.

10. The force of_____ pulls objects toward Earth's center.

machine

move

distance

energy

gravity

motion

speed

force

changing

work

ACTIVITY

Force and Work

Find Out How do forces cause simple machines to work?

What to Do For each picture, write whether a push, a pull or both is acting on the simple machine. Then, describe the work that is being done.

	Force	Work
Bicycle being pedaled	*Feet push on pedals*	*Pedals pull chain*
Pulley lifting weight		
Screw being turned		
Ax splitting wood		
Wheelchair going up ramp		

Simple Machines & How They Work
Simple Machines

The Big QUESTION *How do simple machines make work easier for people?*

Media Resources
Videotape
Grade 3 • Physical Science

Laserdisc
Grade 3 • Physical • Side 1
SAMPLE
Frame 12345

① Introduce

Lesson Background A machine is a device that does work. It takes energy input as a force and acts with the amount of force supplied. A machine cannot produce more energy than is provided to it.

Activate Prior Knowledge Have students call out machines they use or see used on a daily basis. Record their ideas on the chalkboard. Were any "simple" machines (in physical science terms) mentioned? Examples might include: *a knife, wedge, pulley, doorknob, seesaw, needle.* Next, tell students that such tools as rakes, shovels and axes are called simple machines in physical science terms.

② Teach

Before Viewing the Video
For Discussion
Explain to students that there are six types of simple machines upon which all other machines are based, no matter how complicated: the lever, wheel and axle, pulley, inclined plane, wedge and screw. Although the lever, inclined plane and wedge are highlighted in this lesson, you may want to write the names of all six simple machines on the chalkboard. Show the video.

After Viewing the Video
Thinking Critically
1. Ask students to classify the different kinds of simple machines they saw in the video. Which were levers? Inclined planes? Wedges?
2. Ask students what they think is the most common form of simple machine.

Using Video Notes
Distribute copies of Page 104 to students. Show the video again, and tell students to pay close attention to the different types of levers. As they watch, have students label each type of lever on their worksheet as a first-, second- or third-class lever.

Objectives
- explain how an inclined plane works
- explain how a lever works
- describe how other simple machines work

Vocabulary
inclined plane, lever, fulcrum, load, effort

3 Close

Checkpoint

- How does an inclined plane work? (It reduces force by spreading the load out over a greater distance.)
- How do levers work? (A lever works by lifting a load using a rod and fulcrum. Typically, force is applied to one end of the rod to lift the load at the other end.)
- How do other simple machines work? (A wedge is a machine that splits things apart; in a wheel and axle machine, force is applied through a rope or cable to two wheels (one large and one small) that either reduce or increase the amount of force; a pulley spreads a lifting load over many axles; a screw is an inclined plane that spreads a load over its many threads.)

Activity

What to Expect Students will construct a model of a screw and compare it to an inclined plane.

Distribute Page 105 to students and supply the appropriate materials. Encourage students to combine screws and bolts with nuts. Have them answer the questions as they work through the activity.

Alternate Activity

Find Out What kind of simple machines are in our school?

What to Do

1. Write *inclined plane, lever, wedge, pulley, wheel and axle* and *screw* on the chalkboard for students to copy into their notebooks.

2. Take students on a walking tour of the school (especially visiting the playground) and have them look for all the ordinary, everyday places in which we find simple machines.

3. Have students make a chart of the various simple machines they found in the school.

Conclusion

Which type of simple machine is most common in the school?

Was one type of simple machine used more often than another in certain rooms or situations?

VIDEO NOTES

Simple Machines

Directions Watch the video. Label each of the pictures below as first-, second- or third-class levers.

1. _____ -class lever

2. _____ -class lever

3. _____ -class lever

4. _____ -class lever

ACTIVITY

Simple Machines

Find Out A screw is actually a "rolled-up" inclined plane. In this activity, you will construct a model of a screw.

What You Need meterstick, paper, scissors, pencil, wide-tip marker, different size bolts and screws, screws with varying threads, nuts

What to Do

1. Use the meterstick as a straight edge. Draw a diagonal line across the paper with the marker.

2. Cut the paper in two along the edge of the marked line. Save the piece with the line as your good piece.

3. Now you're going to wrap the paper around the pencil. To do this, place the pencil against the paper on a flat surface, with the pointed end of the paper facing away from the pencil, and the lined side of the paper facing down.

4. Slowly wrap the paper around the pencil, keeping the space between the marked line even.

A. What shape resulted from cutting the paper along the diagonal line?

_____

B. What type of simple machine do these pieces of cut paper resemble?

_____

C. Where in your home would you expect to find a screw?

_____

D. Examine several screws and bolts and make a list of how they differ.

_____

_____

E. What force keeps the screw from twisting back out of the nut?

_____

Electricity
Electrical Energy

The Big QUESTION *How is electrical energy changed to other forms of energy?*

Media Resources
Videotape
Grade 3 • Physical Science

Laserdisc
Grade 3 • Physical • Side 1
SAMPLE
Frame 12345

❶ Introduce

Lesson Background Electricity is one of the most important forms of energy. We cannot see, hear or smell electricity, but it is converted into light, heat and mechanical energy that we use for myriad applications. Most of the electricity that we use daily consists of tiny particles called electrons that flow in an electric current. An electric current travels over one or more paths called an electric circuit — a simple loop containing (1) a source of electricity such as a battery, (2) an electric device such as a light bulb, and (3) a conductor between the source and the bulb, such as two wires. One of the wires conducts electricity from the battery to the bulb while the other completes the loop by conducting electricity from the bulb back to the battery. Televisions and computers operate on much more complicated circuits. Electrical energy may be derived from chemical reactions (batteries) or generators fueled by coal, oil, gas, nuclear power or running water. Materials such as aluminum, copper and silver encourage the flow of electrons and thus are good conductors of electricity. Materials such as glass, plastic and rubber inhibit the flow of electrons and are thus nonconductors, or insulators.

Activate Prior Knowledge Ask students to define energy. Record their various ideas on the chalkboard.

Objectives

• define electrical energy

• describe how electrical energy is made and moved

• explain how electrical energy does useful work

❷ Teach

Before Viewing the Video
For Discussion
Challenge students to pretend that a visitor from the year 1000 has just stepped into their classroom. How would they explain the overhead lights, the air conditioning and the computer? What other things that use electricity would baffle the visitor? Write their responses on the chalkboard. Then, turn students' attention to the video. Tell them to note the various forms of energy that electrical energy can become.

Vocabulary

electrical energy, circuit, electric current, conductors, nonconductors

After Viewing the Video
Thinking Critically
Ask students to recall that electrical energy is converted into light, heat and mechanical energy for our use. Review the list of electrical devices on the chalkboard and tell students to identify which type of energy is being used.

Using Video Notes
Distribute copies of Page 108 to students. Have them watch the video once again and answer the questions on their worksheet.

③ Close

Checkpoint

- What is electrical energy? (Electricity is energy that flows in a current through wires to be converted into other types of energy to do work.)

- How is electrical energy made and moved? (Electrical energy is produced by chemical reactions, as in batteries, or by generators that run on coal, oil, gas, nuclear power or water. It flows as a current through conductors.)

- How is electrical energy converted into useful work? (Electrical energy is converted into heat, light or mechanical energy.)

Activity
What to Expect Students will order the steps that bring electricity to our homes.

Distribute Page 109 to students and ask students to chart this path on their Activity pages. Emphasize that hydroelectricity is only one method of producing electric energy.

Alternate Activity
Find Out What safety procedures should be followed in cases of downed power lines?

What to Do
Invite a power company lineman or school maintenance person to explain why downed power lines are dangerous and what students should do in such situations. Have students design and produce a safety brochure to illustrate what to do when one encounters downed power lines.

FUN FACT
The electricity in your home is dangerous, and it's only about 110 volts. The South American electric eel can generate a lethal 500 volts!

ACROSS THE CURRICULUM
Health
Because people do not have to work as hard as they did before electricity and modern appliances were invented, they need less food and more recreational exercise. Have students compare the number of calories used by a typical farmer in the early 1800s with those used by an office worker today.

Health
Electrical energy is extensively used in medicine. Have students research how electronic devices such as the defibrillator and the pacemaker help people with certain heart problems lead normal lives.

VIDEO NOTES

Electrical Energy

Directions Watch the video and listen carefully for clues. Fill in the blanks with the appropriate response.

1. A _____ is a material that allows an electric current to flow through easily.

2. The pathway electricity moves through is called a _____ .

3. _____ produce electricity from chemicals that react inside them.

4. Electrical energy is converted into _____ to make your oven work.

5. Electrical energy is converted into _____ in a flashlight.

6. Electrical energy is converted into _____ to run the motor in your dishwasher.

7. _____ are materials such as plastic or rubber that do not conduct electric current very well.

> conductor
>
> heat energy
>
> circuit
>
> mechanical energy
>
> batteries
>
> nonconductors
>
> light energy

ACTIVITY

Electrical Energy

Find Out What is one way that electricity is produced and brought to our homes?

What to Do

Order the steps that electricity follows to light up a light bulb. Then, draw one way that electricity is produced nearby and comes to your home.

_____ A thin wire in the light bulb heats up. Electrical energy turns into heat energy.

_____ Electricity flows as current through a continuous circuit of power lines.

_____ Turning on a light switch completes the circuit inside your home, allowing electrical energy to flow freely.

_____ Electrical energy flows through wires, or conductors, into your home.

_____ The thin wire gets so hot that it glows and gives off light. Heat energy turns into light energy.

_____ Running water fuels a generator.

Electricity
Electromagnets

The Big QUESTION *How does electrical energy make magnets more powerful?*

Media Resources
Videotape
Grade 3 ● Physical Science

Laserdisc
Grade 3 ● Physical ● Side 1
|||||| SAMPLE ||||||
Frame 12345

Objectives
- define magnetism
- demonstrate how an electric current can make a magnet
- describe how we use electromagnets

Vocabulary
magnetism, electromagnet

❶ Introduce

Lesson Background An electromagnet is a temporary magnet that can be turned on or off by an electric charge. Most electromagnets consist of wire wound around a metal (iron, nickel, cobalt) core. When an electric charge is introduced, the metal becomes magnetized. The strength of the magnet can be increased by either increasing the electric charge or by increasing the number of wire coils around the core. Electromagnets are used in industrial applications to pick up large pieces of metal and also in many devices we use every day — televisions, VCR's, washing machines, hair dryers and electric doorbells.

Activate Prior Knowledge Ask students to explain what magnets do. What kinds of materials do magnets attract? (Metals. However, not all metals are attracted to magnets.) Why do you think that magnets are studied in the same lesson as electricity? (They are both types of energy.)

❷ Teach

Before Viewing the Video
For Discussion
Ask students to describe what it feels like to pull a nail or a paper clip from a magnet. (Students should indicate that a magnet "holds on" to the objects.) What makes one magnet stronger than another magnet? Show the video.

After Viewing the Video
Thinking Critically
1. Ask students to think of different ways to use a magnet that could be turned on and off. Write their responses on the chalkboard.

2. How are electromagnets made stronger? (by increasing the number of wire coils)

Using Video Notes
Distribute copies of Page 112 to students. Have them watch the video once again. This time, have the students watch the video for different applications of electromagnets. Encourage them to add their own applications on their worksheet. (Student responses should reflect understanding of what magnets can affect and where they might be useful.)

③ Close

Checkpoint

- What is magnetism? (the ability of a magnet to exert force)
- How can electricity make a magnet? (by adding electric energy to iron)
- How do we use electromagnets? (televisions, scrap metal collectors, hair dryers, etc.)

Activity

What to Expect Students will construct a weak electromagnet and predict ways to increase its strength.

Distribute Page 113 to groups of three to four students. Help them follow directions to make their own electromagnets.

Alternate Activity

Find Out How is electricity and magnetism related?

What You Need 60-cm copper wire, 6-volt dry cell, stiff cardboard sheet, iron filings

What to Do

1. Cut a hole in the cardboard sheet and run the wire through the cardboard.
2. Connect each end of the wire to each end of the dry cell.
3. Sprinkle iron filings on the cardboard sheet and hold level. The filings will form a series of circles around the cord.

Conclusion

Why are the iron filings affected by the electric cord? (The electric current in the wire forms a magnetic field.)

FUN FACT

Scientists estimate that the magnetic North Pole drifts approximately 10 km per year.

ACROSS THE CURRICULUM

Social Studies

Invite students to research the English physician William Gilbert, also know as William of Colchester. What was his contribution to our present understanding of electricity? What were the clothes, customs and political conditions in England during the 1600s?

VIDEO NOTES

Electromagnets

Directions Watch the video and look for examples of electromagnet uses. List five things that use electromagnets:

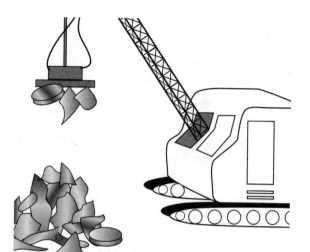

1. _____

2. _____

3. _____

4. _____

5. _____

How would you use a big magnet that could be turned on and off? List and illustrate your ideas.

ACTIVITY
Electromagnets

Find Out How are electricity and magnetism related?

What You Need one meter of #22 copper wire, wire leads, one 6-volt dry cell, two steel nails (at least 10 cm long), ten paper clips

What to Do

1. Leaving about 20 cm of extra wire at the beginning and end, wrap 15 loops of wire around one nail. Be careful not to overlap any loops.

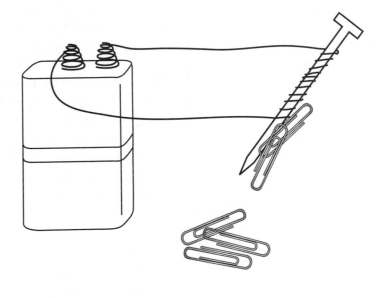

2. Connect one end of the wire to the dry cell. Have one partner hold the other end of the wire to the other end of the dry cell while you carefully move the nail into the pile of paper clips. What happens to the paper clips?_____

_____

3. Pick up as many paper clips as you can. How many do you pick up?_____
Move the nail away from the pile, and then take the end of the wire away from the dry cell. What happens?_____

_____

4. Predict what you have to do to pick up more paper clips. Test your prediction by using the materials to change your setup. Repeat steps 2-3.

Talk About It

Why was the nail able to pick up the paper clips? Why was it not able to pick them up after the wire was disconnected? What could you do to pick up even more paper clips?

Electricity
Static Electricity

The Big QUESTION *What is static electricity?*

Media Resources
Videotape
Grade 3 • Physical Science

Laserdisc
Grade 3 • Physical • Side 1
SAMPLE
Frame 12345

1 Introduce

Lesson Background Static electricity is called "static" because it does not involve a flow, or current, of electrons; it is thus contrasted to "current electricity." It is an imbalance in the number (or build-up) of electrons in matter. The imbalance is often caused by friction created when two things are pulled or peeled apart and electrons are taken by one item from the other — when you pull a comb through your hair, your hair loses electrons and becomes positively charged while the comb gains electrons and becomes negatively charged. The static electricity causes a crackle that you can hear as you comb your hair.

Activate Prior Knowledge Assess students' prior knowledge by asking them the following questions: What does electricity look like? How do you think electricity works? Write these answers on the chalkboard for comparison and contrast to examples shown in the video.

2 Teach

Before Viewing the Video
For Discussion

Review Lessons 1 & 2 of this chapter, emphasizing that electric energy studied so far involves the flow of electricity. Contrast static electricity by explaining that static electricity is actually the build-up of electricity. You may want to blow up a balloon to demonstrate that building pressure eventually results in some type of discharge (the air rushes out of the balloon when it bursts, and the rushing air is like the discharge of built-up electrical energy). Ask students if they can name any examples of static electricity. Show the video.

After Viewing the Video
Thinking Critically

1. Ask students to name the types of static electricity they saw in the video. How do these examples compare to the concepts students named earlier?

2. How dangerous is static electricity? Ask students to name a few of the hazards of static electricity they saw in the video. Make sure they know that common static electricity around the home, such as static cling in socks and blankets, does not start fires.

Objectives
- define electric charges and explain how they build up
- describe how static electrical charges behave
- describe the effects of static electricity

Vocabulary
electric charges, static electricity

Using Video Notes

Distribute copies of Page 116 to students. Give them time to review the statements on their worksheet and then encourage them to look for things that static electricity does and doesn't do as they watch the video once again. This time, have students determine whether each statement on their worksheets is true or false.

③ Close

Checkpoint

- What are electrical charges, and how do they build up? (Electrical charges are electrical forces that exist in all matter. When objects rub together, electric charges build up in both objects.)

- How do static electrical charges behave? (Static electrical charges cause objects to stick together if each object has a different electrical charge; if both objects have the same electrical charge, they will push apart.)

- What are the effects of static electricity? (static cling, sparks, lightning, small shocks)

Activity

What to Expect Students will conduct an experiment with static electricity and explain their results.

Distribute Page 117 to students and provide the necessary supplies. Don't forget to recycle the used cans when the experiment is finished!

Alternate Activity

Find Out Can static electricity attract water?

What You Need water faucet, plastic comb and your hair, or a rod and some wool, dry air conditions

What to Do

1. Turn on the faucet so that the water runs slowly and smoothly without breaking up.

2. Create static electricity by combing your hair with a plastic comb or rubbing a plastic rod with some wool.

   Note: Thin hair tends to generate more static electricity than thick hair.

3. Place the charged object near the running stream of water.

Conclusion

What happened to the stream of water? (The stream of water should be attracted by the comb.) Why? (The stream of water and the comb have opposite charges.)

VIDEO NOTES

Static Electricity

Directions Circle T or F to show whether each statement is true or false.

T F 1. Clothes clinging and sparks flying from static electricity around your house is very dangerous.

T F 2. Static electricity is the build-up of electricity; it does not flow.

T F 3. Lightning is caused by static electricity.

T F 4. Lightning rods get hit by lightning first because they are higher than everything else.

T F 5. A good way to get rid of static electricity is to ground it.

T F 6. Computer technicians need to be careful because static electricity can shock and hurt them.

T F 7. When objects have the same charge, they are attracted to each other.

T F 8. Static electricity only builds up on a television screen when the television is off.

ACTIVITY
Static Electricity

Find Out How does static electricity build up?

What You Need empty aluminum drink can, inflated balloon, your hair

What to Do

1. Put the can on a flat, smooth surface such as the floor or table.
2. Rub the balloon back and forth on your hair or on a wool sweater. Follow the experiment. Record your results in the right-hand column.

Try this...	What happened?
Hold the balloon about 25 mm in front of the can.	
Move the balloon away from the can slowly.	
Move the balloon to the other side of the can.	
Have a can race with someone else.	
Roll the can uphill.	

Talk About It

Why does the can roll? In the race, whose can rolled faster? How could you make your can roll faster? How far can you roll the can before it stops? How could you make it roll farther?

Muscles & Bones Work Together
Muscles and Bones
The Big QUESTION *Why are muscles and bones important?*

Media Resources
Videotape
Grade 3 • Health Science

Laserdisc
Grade 3 • Health • Side 1
SAMPLE
Frame 12345

1 Introduce

Lesson Background The human body is composed of several systems. Two of these systems are the skeletal system and the muscular system. The skeletal system is composed of bone, cartilage, ligaments and tendons; it acts as a support frame for the body and provides protection for internal organs. The muscular system is composed of tissues that cover the bones of the skeleton and make up parts of certain organs. These tissues are connected to tendons that connect them to bones. Muscles are used to move parts of the body, including the actions of some internal organs (the heart, the stomach).

Activate Prior Knowledge Determine student understanding by asking students a series of simple questions: What do bones do? What do muscles do? How do bones and muscles work together?

2 Teach

Before Viewing the Video
For Discussion

Explain that humans use muscles for movement. Challenge students to think of parts of the body that do not have obvious muscles (ears, nose), and ask them if those parts have bones and joints in them. Tell students to think about how muscles and bones work together as they watch the video. Next, use the analogy of a lunchbox protecting chips and a sandwich to tell students that one function of bones is to protect internal organs. Now ask them if they can think of any bone structures that protect important organs (rib cage protects heart and lungs; skull protects brain; the spinal column protects the spinal cord).

After Viewing the Video
Thinking Critically

To assess students' comprehension of skeletal and muscular functions, encourage students to describe the working relationship of bones and muscles: What holds the bones together? (ligaments) What holds the muscles to the bones? (tendons) Why are both the muscles and the bones needed? (Bones provide a strong frame while muscles move the bones.)

Using Video Notes
Distribute copies of Page 120 to students. Have them watch the video again to help them think through their drawings. After viewing the video, have students draw answers on their worksheet.

Objectives
- explain the function of muscles
- explain the purpose of bones
- explain how muscles and bones work together

Vocabulary
- muscular system, calcium, skeleton, skull, ligaments, tendons

3 Close

Checkpoint

- What do your muscles do? (help your body and organs carry out movements)
- What do your bones do? (support the body and protect internal organs)
- How are bones connected? (interlocked by joints and joined by ligaments)

Activity

What to Expect Students will conduct experiments, observe how different muscles have different strengths and functions, and record their findings.

Distribute copies of Page 121 and have students work in pairs. Make sure that students are recording their observations and timing their partners.

Alternate Activity

Find Out How strong is a bird bone?

What You Need paper, tape, ruler

What to Do Birds have many bones that are hollow yet strong. To demonstrate the strength of bird bone design, ask students to:

1. Roll a sheet of paper into a tube approximately 2 cm in diameter. Tape the tube of paper in at least three places. Then repeat, rolling other sheets into various diameters between 3 cm and 5 cm.
2. Stand the tube upright on a table and see how many books can be balanced on the tube of paper. Try different diameters and see which ones work the best.
3. Lay the tube horizontally on a table and see how many books can be placed on the tube.

Conclusions

How do hollow bones help birds? (Birds can fly because their bones are strong and lightweight.)

What happens if the tube of paper is crumpled only slightly? (The tube is weakened and holds less weight.)

What happens when you put books on the tube when it is on its side? (The tube collapses. It is strongest when used in a vertical position.)

FUN FACT

It takes 17 muscles to smile and 43 muscles to frown.

VIDEO NOTES

Muscles and Bones

Directions Draw pictures of how your body would look if it had . . .

no bones

no muscles

no ligaments

no tendons

ACTIVITY
Muscles and Bones

Find Out What happens when muscles contract?

What You Need a watch with a seconds hand

What to Do

1. Pick a partner. While your partner relaxes and flexes muscles, feel the muscle and record its hardness (contracted) or softness (relaxed) in the table to the right.

Muscle	Relaxed	Contracted
Calf		
Shoulder		
Biceps		
Hand		

2. In the table below, record how long your partner can contract his or her muscles.

Muscle	Activity	Time
Calf	Stand on tiptoes	
Shoulder	Lift a heavy book in each hand while spreading your arms out horizontally	
Biceps	Lift a heavy book in each hand near your waist and curl it toward your chest	
Hand	With your whole arm resting on a table, hold a heavy book on your fingertips	

Talk About It How do muscles change when they're contracted? Why? Drawing from your observations, form a hypothesis of which muscles are stronger and why. How many muscles can you feel working in each activity?

Muscles & Bones Work Together
How Muscles Move Bones

The Big QUESTION *How do muscles move bones?*

Media Resources
Videotape
Grade 3 • Health Science

Laserdisc
Grade 3 • Health • Side 1
SAMPLE
Frame 12345

1 Introduce

Lesson Background Muscles are either voluntary or involuntary: voluntary muscles move when a person thinks about moving them, while involuntary muscles move automatically without the person consciously trying to move them. Both voluntary and involuntary muscles are controlled by messages from the brain.

Activate Prior Knowledge Facilitate student understanding by defining voluntary and involuntary as "actions you do on purpose" versus "actions that happen automatically." Ask students to think of things that their bodies do without them having to think about it. (heart beating, blinking, digesting) Contrast this list of actions with an open forum on things we consciously tell our bodies to do. (walk, ride bikes, eat)

2 Teach

Before Viewing the Video
For Discussion

Remind students that muscles usually work as partners, with one relaxing and one contracting. Ask them to describe what happens when certain actions occur. For example, which bones and muscles are working when a person bends his or her arm? (Two main bones are connected at the elbow joint; biceps contract to pull up the forearm and the triceps relax; to straighten the arm, the triceps contract and the biceps relax.) Tell students to watch the video to find out how muscles move bones.

After Viewing the Video
Thinking Critically

1. What happens within your body when you make a fist? (You think about making a fist; your brain sends messages to muscles attached to finger bones; these muscles contract and cause finger bones to bend at the joints.)

2. What happens within your body when you blink? (Your brain senses that your eyes are dry, so it sends a message to the muscles in your eyelids; these muscles contract and cause your eyelids to blink, moistening your eyeballs.)

Using Video Notes

Distribute copies of Page 124 to students. Have them watch the video again, then complete the chart on their worksheet.

Objectives
- explain how muscles work
- compare and contrast voluntary and involuntary muscles

Vocabulary
contracting, voluntary muscles, involuntary muscles

❸ Close

Checkpoint

- What happens within your body when you need to lift a heavy weight? (Your brain sends a message to your muscles and your muscles move your bones.)
- What body parts are working that you don't have to think about? (Digestive system is processing food, heart is beating, eyes are blinking, lungs are breathing and so on.)
- What body parts can you use whenever you want to? (legs to walk/run, face to make expressions, eyes to look around and so on)
- How are voluntary and involuntary muscles the same? (They both contract and relax, and they are used to move parts of the body.) How are they different? (Voluntary muscles can be used consciously; involuntary muscles operate without conscious control.)

Activity

What to Expect By working in pairs, students can explore their control over their own muscles while illustrating which muscles are voluntary and which muscles are involuntary.

Distribute copies of Page 125. Have students follow the steps outlined to examine voluntary and involuntary muscles.

Alternate Activity

Find Out How do muscles work?

What to Do

1. Ask one student to stand in the doorway of the classroom, place the backs of his or her hands against the sides of the door jam and apply pressure upward and outward for at least 30 seconds.
2. Now ask the student to step forward. The student's arms should rise without effort. Explain that the arms rise because the arm muscles are still in a state of contraction.

Note: This exercise may also be done in pairs: one student tries to press the backs of his or her hands outward (palms facing) while his or her partner tries to force them together (for about 30 seconds). When the resistance is released, the student should feel the same effect of weightlessness in his or her arms.

Conclusions

When your partner let go of your arms, what were your muscles doing? (The arm muscles were still contracting and thus automatically lifted the arms.)

FUN FACT

An elephant's trunk contains more than 40,000 muscles. These, along with two fingerlike appendages, allow an elephant to pick up fruits as small as marbles or branches a foot thick. By comparison, there are only 700 muscles in the entire human body!

ACROSS THE CURRICULUM

Physical Education
Ask a local weightlifter to give a brief talk to the class about how he or she builds big muscles. Ask the guest speaker to talk about diet and weightlifting, and how much work it takes to make muscles larger. Emphasize to students that everyone has muscles, even if some are not quite as visible, or developed, as others. You may want to contrast the musculature of other types of athletes, e.g., swimmers, runners and gymnasts.

VIDEO NOTES
How Muscles Move Bones

Directions Watch the video, then list the muscles and bones (and perhaps organs) that the body uses to achieve each action below.

Action	Muscles and Bones
Talking	**Muscles:** face, tongue, heart **Bones:** jawbone
Running	
Doing Homework	
Arm Wrestling	
Sneezing	

ACTIVITY

How Muscles Move Bones

Find Out Which of my muscles are voluntary and which are involuntary?

What to Do

1. Team up into pairs and help each other fill out the chart below.

2. Try to determine whether each muscle is voluntary or involuntary. Some muscles may be both.

Muscle	Voluntary	Involuntary
Biceps *Can you flex your arm?*		
Blood Vessels *Can you move your blood?*		
Eyelids *Can you blink your eyes?*		
Jaw *Can you yawn?*		
Toes *Can you wiggle your toes?*		
Intestines *Can you digest your food?*		
Thigh *Can you move your leg?*		
Diaphragm *Can you take a deep breath?*		
Heart *Can you make your heart beat?*		
Eyes *Can you look around?*		

Muscles & Bones Work Together
The Major Bones and Muscles

The Big QUESTION *Why do we have different kinds of muscles and bones?*

Media Resources
Videotape
Grade 3 • Health Science

Laserdisc
Grade 3 • Health • Side 1

SAMPLE

Frame 12345

Objectives

- relate the shape and size of a bone to its function
- relate the type of muscle to its function
- describe the structure of a bone

Vocabulary

cardiac muscle, skeletal muscles, smooth muscles

❶ Introduce

Lesson Background Human bones are cylindrical shapes that are hard on the outside with a yellow or red spongy substance called marrow in the middle. The four kinds of bone types are long, short, flat and irregular.

There are three types of muscles in the human body: cardiac, skeletal and smooth. Cardiac muscle, an involuntary muscle, is found only in the heart and is striated, or striped, in appearance. Skeletal muscles are voluntary muscles found in the legs, arms, abdomen, face, neck and chest. They are attached to the bones by tendons and always work in pairs: one muscle or set of muscles will contract while the other is relaxed. Skeletal muscle fibers are also striated. Smooth muscles are involuntary muscles found in internal organs (stomach, intestines and bladder) and aid blood flow in blood vessels. Smooth muscles, as their name indicates, are smooth in appearance and have no striations.

Activate Prior Knowledge Initiate student understanding of bone types by asking them to name bones that are long (legs, arms), bones that are short (hands, feet), bones that don't move (flat bones, including ribs and skull) and irregularly shaped bones (face bones, middle ear bones, backbone). To introduce muscle types, ask students to name things inside and outside of their bodies that move. (arms, legs, eyeballs, heart, stomach, blood)

❷ Teach
Before Viewing the Video
For Discussion

Remind students that muscles and bones work together. Ask them to bend one arm and to imagine the arm as two long bones connected by muscles that pull the forearm up into an angle. Now ask them to straighten their arms. Which muscles performed this task? How are the two bones connected? (Muscles can only pull, not push. The biceps contract to pull the forearm up while the triceps relax. The opposite happens when they straighten out their arms. Bones are connected in a joint.) Tell students to pay close attention to the types of muscles and bones shown in the video.

After Viewing the Video
Thinking Critically

1. What are the four kinds of bones? (long, short, flat, irregular)

2. What are the three kinds of muscles? (skeletal, smooth, cardiac)

3. Describe the structure of a bone. (cylindrical, hard on the outside with a spongy substance on the inside)

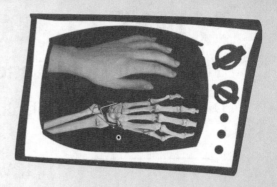

Using Video Notes

Distribute copies of Page 128 to students. Have them watch the video once again and listen for clues that will help them complete their worksheet.

❸ Close

Checkpoint

- What kinds of bones do you use for walking? (long bones)
  What kinds of bones do you use for hearing? (irregular bones)
  What kinds of bones don't move? (flat bones)
  What kinds of bones do you use for writing? (short bones)

- What kinds of muscles do you use for walking? (skeletal muscles) What kind of muscle does your heart use to pump blood? (cardiac muscle) What kinds of muscles does your body use to digest food? (smooth muscles)

- What do bones look like on the inside? (They have a space filled with a soft substance called marrow that is either yellow or red.)

Activity

What to Expect Students will describe bone function and location by creating their own names for bones.

Distribute copies of Page 129. Have students follow the steps outlined to label their own personalized skeleton.

Alternate Activity

Find Out Why is the skull such a great design?

What You Need half a dozen eggs, one bowl

What to Do

1. Ask a volunteer to take an egg with their forefinger and thumb and try to break the egg by squeezing it along the vertical axis. Make sure they have taken off any rings.

2. Have the students try to break an egg along its horizontal axis.

Students are not able to break the egg because the yolk is carrying much of the force that is exerted on the egg. The dome's shape is strong because it spreads out weight over a large area. You may want to add that the tubular shape of bones is similar to columns that architects use for support in houses and buildings.

Conclusions

What body part is similar in shape to the egg? (the skull)

What does the skull protect? (the brain)

FUN FACT

Your *femur*, or thighbone, the largest bone in your body, is pound-for-pound stronger than an equal amount of solid steel.

ACROSS THE CURRICULUM

Art

Describe the structure of bones to the students, including information about marrow. Working from your description, have students illustrate what bones look like both inside and outside. Have them label their drawing with descriptive words. For example, would they use the word *hard* or *shiny* to describe the outside of a bone? You might also help students create papier-maché models of bones.

Language Arts

Use a dictionary to look up words that start with *cardi-* and find out how many words have been invented based on this old Greek word meaning "heart."

VIDEO NOTES
The Major Bones and Muscles

Directions Fill in the blanks by choosing the best word from the word bank.

long

short

irregular

flat

voluntary

involuntary

cardiac

smooth

skeletal

1. When I eat, I move my jawbone. The jawbone is a(n) _____ bone.

2. Even when I'm sleeping, my heart pumps blood through my body. The heart is a(n) _____ muscle.

3. My heart and lungs are protected inside my ribcage, which does not move. My ribcage is made up of _____ bones.

4. _____ bones support my arms and my legs.

5. My fingers do very detailed work, like drawing and writing. They have a lot of _____ bones.

6. After I've eaten dinner, my stomach starts to digest the food. The stomach has _____ muscles in it.

7. When I make a muscle with my biceps, I can see a(n) _____ muscle.

8. The only place to find _____ muscle is in the heart.

9. If I wanted to jump up and down right now, I could. I'd be using _____ muscles.

ACTIVITY

The Major Bones and Muscles

Find Out Where are bones located in the human body? What do they do?

What to Do Invent your own names for bones using your common-sense knowledge of what they do and where they are found in the human body. Label the skeleton below with your own code names. For example, you might call the *scapula* the *shoulder bone* or even the *backpack holder*.

Note: aka means *"also known as"* . . .

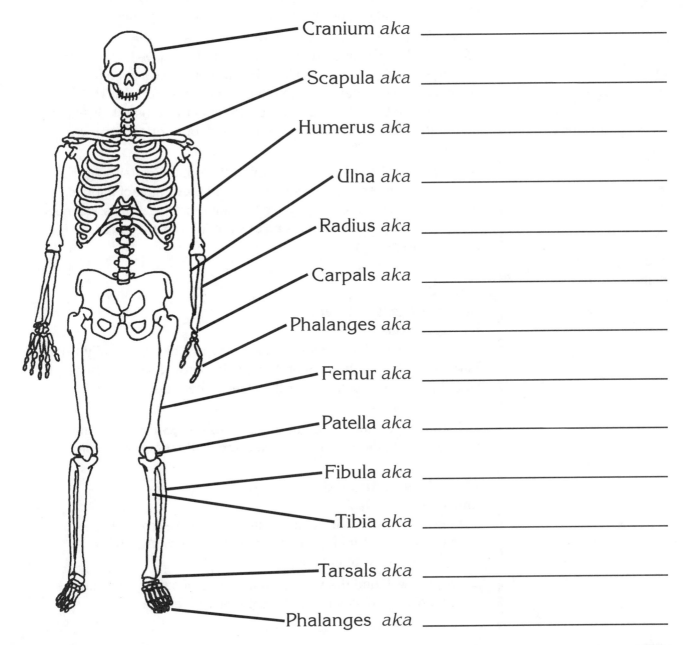

Cranium *aka* _____

Scapula *aka* _____

Humerus *aka* _____

Ulna *aka* _____

Radius *aka* _____

Carpals *aka* _____

Phalanges *aka* _____

Femur *aka* _____

Patella *aka* _____

Fibula *aka* _____

Tibia *aka* _____

Tarsals *aka* _____

Phalanges *aka* _____

Personal & Community Health
Avoiding Injuries

The Big QUESTION *How can injuries be prevented?*

Media Resources
Videotape
Grade 3 ● Health Science

Laserdisc
Grade 3 ● Health ● Side 1
SAMPLE
Frame 12345

Objectives
- describe safety equipment that keeps you safe at work and play
- explain how seat belts keep you safe

Vocabulary
injuries, helmet, seat belts

❶ Introduce

Lesson Background Safety equipment prevents injuries at work and play. Each sport (bicycling, soccer, hockey, basketball, baseball, in-line skating and so on) has specific safety gear, but basic equipment can be used in many activities (helmets, shin and wrist guards, knee and elbow pads). Other sports require specific gear (football requires more padding and mouth guards; basketball players should wear high-top shoes to protect against ankle injuries). Safety gear for some work environments, such as hard hats, goggles and gloves, is of utmost importance.

Seat belts have been proven to save lives in car crashes. Currently, 49 states plus Washington D.C. and Puerto Rico have laws requiring car drivers and passengers to wear seat belts. Laws specific to protecting children are also in effect; as of 1995, the NHTSA estimated that child restraints are 71% effective if used correctly, although actual effectiveness (due to improper operation) was estimated to be 59%.

Activate Prior Knowledge Determine student understanding of the connection between safety equipment and prevention of injuries. Ask students about specific types of equipment (seat belts, helmets, goggles) and elicit responses as to what injuries they are designed to prevent.

❷ Teach
Before Viewing the Video
For Discussion
Ask students to think of sports they play in which they might get hurt. Write their responses on the chalkboard. Next, ask them what kinds of protective gear they wear when they engage in those activities.

Tell students to note different types of safety gear for different activities (including work) that they see as they watch the video.

After Viewing the Video
Thinking Critically
1. For each item of safety gear listed on the chalkboard, ask students what that item does to protect them. Add their responses next to each item.

2. As they watched the video, which types of safety gear did they recognize, and which types were new to them?

Using Video Notes

Distribute copies of Page 132 to students. Have them watch the video again. This time, have students fill in the blanks on their worksheet with the safety equipment each character should be wearing.

③ Close

Checkpoint

- What safety equipment should be worn when you are skating? (wrist guards, knee and elbow pads, helmet)
- What safety equipment should be worn when you are working in a chemistry lab? (safety glasses, gloves, apron)
- How do seat belts keep you safe? (minimize human impact and prevent ejection from car)

Activity

What to Expect Students will conduct a safety inspection of their playground, identify potential risks and recommend improvements.

Distribute copies of Page 133. Have students follow the steps outlined to conduct a safety inspection of their playground.

Alternate Activity

Find Out What are the basic safe operating requirements for an automobile?

What to Do

1. Tell students to conduct a safety inspection of their family car or an adult friend's car. Warn them to use extreme caution during their inspection and to complete this activity only with the help of an adult. Emphasize that under no circumstances should any student operate any part of the car.

2. Help each student complete an inspection chart similar to the one used for the playground inspection. The automobile inspection chart should include checkpoints for properly working headlights, high beams, taillights, brake lights, right blinker, left blinker, parking lights, seat belts, windshield wipers, tire pressure and horn.

Conclusions

Did your automobile pass your inspection?

How could you improve the safety of your automobile?

FUN FACT

The National Highway Transportation Safety Administration (NHTSA) estimates that seat belts save approximately 9500 lives in America each year. As of 1999, only New Hampshire does not have a seat belt law for adults.

ACROSS THE CURRICULUM

Technology

Ask students to write down a list of automobile hazards. Unsafe drivers? Speeding? Hard/sharp edges? Ask them to design and draw a car with new safety features.

VIDEO NOTES

Avoiding Injuries

Directions Fill in the blanks with the safety equipment that each person should be wearing.

In-Line Skater

_____

_____

_____

Lacrosse Player

_____

_____

_____

knee pads

gloves

hard hat

mouth guard

work gloves

back brace

elbow pads

goggles

wrist guards

helmet

lab coat

ear protectors

face guard

Lab Technician

_____

_____

Construction Worker

_____

_____

_____

ACTIVITY

Avoiding Injuries

Find Out Is our playground safe?

What to Do

1. Inspect your playground for conditions that may cause accidents.

2. Write down designs that are safe and write down places where you could be hurt.

Playground Conditions	Yes	No
Is there any rusted metal?	——	——
Are there any sharp edges?	——	——
Are there any edges or hooks that could pinch?	——	——
Are all nuts, bolts and clamps capped off?	——	——
Is the ground a soft surface (sand, bark, gravel or rubber)?	——	——
Is there enough space for several people to play?	——	——
Are all openings (hanging rings, stair rungs, spaces between railings) either less than 13 cm or more than 25 cm?	——	——
Are all nuts, bolts and clamps tightened?	——	——

Comments

Talk About It What types of safety considerations were put into the design of our playground? How can we improve the safety of our playground?

Personal & Community Health
Basic First Aid

The Big QUESTION *What are basic first aid procedures?*

Media Resources
Videotape
Grade 3 • Health Science

Laserdisc
Grade 3 • Health • Side 2
||||| SAMPLE ||||
Frame 12345

Objectives
- define first aid
- explain how to give first aid treatment for scrapes, nosebleeds, cuts and sprains

Vocabulary
first aid, sprain, nosebleed

1 Introduce

Lesson Background First aid is medical care given immediately to a victim of an accident, sudden illness or other medical emergency. Proper first aid can save a person's life and prevent the development of related medical problems. The first step of emergency care is to assess the situation: if you are confused or unsure of yourself, do *not* try to give treatment. Instead, call for help as quickly as possible. Basic first aid procedures covered in this lesson include how to treat cuts and scrapes, nosebleeds and sprains. Emphasize to students that it is very important to find an adult to help.

Activate Prior Knowledge Describe situations of varying seriousness and ask students to identify each incident as a minor accident or an emergency. For example, describe an incident involving a scraped knee, then recount an accident involving a broken arm.

2 Teach

Before Viewing the Video
For Discussion
Draw columns on the chalkboard. Label one "Cuts/Scrapes," a second "Nosebleed" and a third "Sprains." Ask students to name some things they could do to care for each occurrence. As students call out their ideas, list them in the appropriate columns. Tell students that, as they watch the video, they should look for other first aid treatments.

After Viewing the Video
Thinking Critically
Have students recall the accidents and treatments they saw in the video. Add these to the appropriate column on the chalkboard. Do they want to change anything that was written before watching the video?

Using Video Notes
Distribute copies of Page 136 to students. Have them watch the video again. This time, have students outline steps to follow on their worksheets for each incident named.

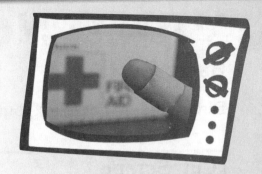

③ Close

Checkpoint

- What is first aid? (immediate care given to the victim of an accident)
- How should a cut or scrape be treated? (clean cut, apply antibiotic ointment, cover with bandage)
- How should a sprain be treated? (rest the joint, apply an ice pack, and elevate above the heart)
- How should a nosebleed be treated? (Sit down, lean slightly forward and pinch nostrils shut for five to ten minutes until bleeding stops; if bleeding continues, put cold cloths on nose.)

Activity

What to Expect Students will identify proper actions to take in specific situations.

Distribute copies of Page 137. Ask students to fill in the blanks with the incidents that match the actions described in the boxes.

FUN FACT

The National Safety Council reports that bicycles, stairs and doors, in that order, cause more accidents in the home than any other objects.

ACROSS THE CURRICULUM

Careers/Social Studies
Using the key words *American Red Cross*, students can use an encyclopedia or the Internet to research the origins of the nation's largest supplier of blood, plasma and tissue. Ask students to write short sentences describing its beginnings, its activities and career opportunities.

VIDEO NOTES

Basic First Aid

Directions In each box, outline the steps to follow in giving first aid treatment.

Cut/Scrape

Nosebleed

Sprain

ACTIVITY

Basic First Aid

Find Out What should you do when someone gets hurt?

When should you . . .

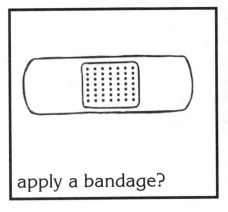

apply a bandage?

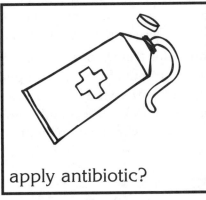

apply antibiotic?

apply a cold pack?

_____

_____

_____

clean an injury?

find an adult?

call 911?

_____

_____

_____

Personal & Community Health
Emergency Care

The Big QUESTION *What can be done to help someone who is injured?*

Media Resources
Videotape
Grade 3 • Health Science

Laserdisc
Grade 3 • Health • Side 2
SAMPLE
Frame 12345

1 Introduce

Lesson Background When possible, the best way to handle a medical emergency is to get help by dialing 911. When making an emergency telephone call, you should be ready to describe the nature of the victim's injury and the exact location and telephone number where the accident occurred. Ideally, you should be prepared to write down instructions that the operator may give.

Activate Prior Knowledge Ask students to recount different incidents that qualify as an emergency versus a minor event. Has any student ever called 911? What happened? With the students' help, begin a list on the chalkboard of things that can be done to help someone who has been injured.

2 Teach

Before Viewing the Video
For Discussion
Remind students to call 911 for serious medical emergencies. Then, ask students if they can think of anyone else that they should call. Show the video with a reminder to students to look for basic steps to follow in case of an emergency.

After Viewing the Video
Thinking Critically
1. Ask students to list the typical things that happen on a trip to the emergency room. As they call out these steps, add them to the list on the chalkboard.

2. Ask students how to be prepared when making a 911 call. (know exact location of victim and the nature of the victim's injury or illness, have pen and paper to write down instructions)

Using Video Notes
Distribute copies of Page 140 to students. Have them watch the video once again, then answer the questions on their worksheets.

Objectives
- list rules to follow during an emergency
- specify steps to follow when making an emergency telephone call
- summarize what happens during an emergency room visit

Vocabulary
emergency, cast, stitches

3 Close

Checkpoint

- What rules should be followed in case of an emergency? (stay calm, dial 911, notify an adult)
- What questions will be asked during an emergency telephone call? (name, address, telephone number, nature of accident)
- What happens during a visit to the emergency room? (ambulance takes victim to the hospital as quickly as possible, victim registers at front desk and is then sent to "triage," where he or she waits for a doctor)

Activity

What to Expect Students will identify efficient steps to follow in handling an emergency and dramatize these steps for the class.

Distribute copies of Page 141. Assign students to groups of four or five members. Give each group an emergency situation such as an accidental poisoning, a broken bone, a large cut or a fire in the home. Instruct students to complete the drawings first, using them as a guide, or storyboard, to their skit.

Alternate Activity

Find Out Who do I call when an accident occurs?

What You Need posterboard, pens, telephone book

What to Do

1. With their parents' help, have students write down their emergency phone numbers (home phone number, 911, police department, fire department, poison control center, doctor's office, parents' work phone numbers, other adult's phone number, veterinarian).

2. Let students know that the front of the telephone book will have some of these numbers.

3. After the list is made, students should place it somewhere at home where everyone in the house can easily access it.

Conclusions

When should you call 911? (in an emergency only)

When should you call the police? (if you see something bad happening, e.g., someone breaking into a neighbor's house)

When should you call the Poison Control Center? (when someone has swallowed something that may be poisonous or has gotten something in his or her eyes that may be harmful)

FUN FACT

Firefighters perform the most dangerous job. They must be busy, too. In the United States, a house catches fire every 45 seconds.

ACROSS THE CURRICULUM

Careers

Have an emergency room nurse or doctor, a firefighter, or a paramedic visit the class to tell students what happens during a trip to the emergency room.

VIDEO NOTES

Emergency Care

Directions Imagine that you are at a friend's house. Your friend is climbing a tree and falls. He or she is very hurt. Order the five basic steps you would take in this emergency situation.

_____ Tell your friend that you are calling 911.

_____ Call 911.

_____ Put a blanket over your friend to keep him or her warm.

_____ Find out the address of your friend's house.

_____ Tell the operator "This is an emergency."

Your friend is zoomed safely to the hospital in a:

a. school bus

b. station wagon

c. ambulance

d. race car

Your friend has a very deep cut. The doctor pulls out a special sewing thread and stitches the cut. This helps to:

a. stop the bleeding

b. pull the cut back together

c. encourage healing

d. all of the above

The doctor also says that your friend's arm is broken! The doctor will fix it by:

a. waving a magic wand

b. applying a lotion

c. putting it in a cast

d. giving a shot

ACTIVITY

Emergency Care

Find Out How would you care for someone in an emergency?

What to Do

1. Your teacher will assign an emergency that your group must handle.

2. Draw and describe the emergency and what your group would do to deal with the problem.

3. Using your group's drawings, act out what you have drawn and written.

Nutrition
Nutrients in the Basic Food Groups
The Big QUESTION *Why are some foods better for you than others?*

Media Resources
Videotape
Grade 3 • Health Science

Laserdisc
Grade 3 • Health • Side 2
SAMPLE
Frame 12345

Objectives
- demonstrate how nutrients contribute to good health
- give examples of nutrients in various foods

Vocabulary
nutrient, carbohydrate, protein, fat, vitamin, mineral

❶ Introduce

Lesson Background Scientists have specified six nutrients (carbohydrates, proteins, fats, vitamins, minerals, water) and related them to the five food groups (grains, vegetables, fruit, dairy, meat) to form the USDA Food Guide Pyramid. This guide suggests dietary quantities that the average American should consume to maintain a balanced diet. Additionally, the FDA now requires that all packaged foods produced in the United States include a nutritional food label that lists "Nutrition Facts." These labels report the food's amount of specific nutrients and vitamins for which the FDA has determined a Required Daily Allowance (RDA). The "Nutrition Facts" label also must report the "% Daily Value" of the RDA for each element the food contains.

Activate Prior Knowledge Draw five columns on the chalkboard and label each column with the name of a food group. Ask students about their most recent meals (What did you eat for breakfast? What did you eat for dinner last night?). Assign five students at a time to stand by each column on the board, and direct them to write down each food item that belongs in their group.

❷ Teach
Before Viewing the Video
For Discussion

Explain to students that people need nutrients to stay healthy. Write all six nutrients on the board. Ask students to name foods they know are high in one of these nutrients; if they don't know for sure, provide examples: *peanut butter is high in protein, ice cream is high in fat, bread is high in carbohydrates, carrots are high in vitamin A, milk is high in calcium.* Tell students to watch the video for more foods that have high nutrient values.

After Viewing the Video
Thinking Critically

1. Have students recount the nutritious foods they saw in the video. Add these foods to the appropriate chalkboard list.
2. Are there any foods that are high in more than one kind of nutrient? (Broccoli has a lot of vitamins C and A and calcium; sardines have a lot of vitamin A, calcium and iron and so on)

Using Video Notes
Distribute copies of Page 144 to students. Have them watch the video again. This time, have the students complete the chart on their worksheets.

3 Close

Checkpoint

- What are nutrients? (substances the body needs to stay healthy) What are the six kinds of nutrients? (carbohydrates, proteins, fats, vitamins, minerals, water)
- What kinds of nutrients do bread, cereal, rice and pasta give your body? (carbohydrates, fats)
- What kinds of nutrients do fruits provide? (vitamins C and A)
- What kinds of nutrients do vegetables give your body? (vitamins C and A and calcium)
- What kinds of nutrients do meat, poultry, fish, dried beans, eggs and nuts give your body? (protein, iron, vitamin A, fat)
- What kinds of nutrients do milk, yogurt and cheese give your body? (vitamin D, calcium, fat)

Activity

What to Expect Students will list daily values from nutritional labels of three foods that they normally eat and add up each figure to show how much of the recommended daily allowance of nutrients they would receive if these were the only foods they ate. Referencing the food pyramid, they will name which food groups they should add to eat a healthy, balanced diet.

Distribute copies of Page 145 and tell students to complete each column by referring to their nutritional labels.

Note: Before students can complete this activity, they must bring in three nutritional labels from foods that they normally eat at home.

FUN FACT

Did you know that what an average person eats over a lifetime weighs as much as 100 cows or 20 rhinos? That's nearly 41 metric tons. Gulp . . .

ACROSS THE CURRICULUM

Physical Education
Tell students they are about to burn some calories, and organize them into two teams of about 15 members each. Give each team one grocery bag, six grain group foods, three vegetables, two fruits, two dairy products and two meat items (or have students make these items out of construction paper). In the gym or on the playground, set up the teams for a relay race. The first runner's grocery bag should have only one item in it. At the end of the lap, he or she should hand off the grocery bag to the next teammate, who will put his or her food item in it and run the next leg of the race. The last runner of each team should be holding all of the team's food items, equal to one day's fully balanced diet.

VIDEO NOTES

Nutrients in the Basic Food Groups

Directions List the foods you see in the video under the correct food group.

bread, cereal, rice and pasta	vegetables	fruits	milk, yogurt and cheese	meat, poultry, fish, beans, nuts and eggs

ACTIVITY

Nutrients in the Basic Food Groups

Find Out Can you get enough of the right kinds of nutrients from just a few kinds of food?

What You Need With your parents' permission, bring in three "Nutrition Facts" labels from home to class.

What to Do

1. Ask an adult family member to help you cut the labels off three food products. Make sure it's all right with him or her before you cut.

2. Based on one serving, write down the percentage amount of fat, carbohydrates, protein, iron, calcium, vitamin C and vitamin A in each of your foods.

3. Total the three percentages in the last column.

Nutrients	Item One	Item Two	Item Three	% Daily Value
Total Fat	%	%	%	%
Total Carbohydrate	g	g	g	g
Protein	%	%	%	%
Iron	%	%	%	%
Calcium	%	%	%	%
Vitamin C	%	%	%	%
Vitamin A	%	%	%	%

Talk About It Did your foods provide 100 percent of each nutrient? If not, what types of foods (from which food groups) should you add?

Nutrition
The Benefits of Good Nutrition

The Big QUESTION *Why is it important to eat a balanced diet?*

Media Resources
Videotape
Grade 3 • Health Science

Laserdisc
Grade 3 • Health • Side 2
SAMPLE
Frame 12345

Objectives
- describe a balanced diet
- give examples of how nutrients help your body

Vocabulary
diet

1 Introduce

Lesson Background The body breaks down food and uses the nutrients to keep the body healthy and strong. Scientists have proven that specific vitamins and minerals afford specific benefits for human health (carbohydrates provide fuel for our bodies; proteins make up every muscle and organ; fats contribute energy reserves; vitamin E helps form red blood cells and utilizes vitamins K and A; vitamin K is essential for blood clotting; calcium promotes strong teeth and bones; zinc aids night vision, taste perceptions and wound healing; magnesium prevents tooth decay; water supplies minerals, carries away waste and cools the body). The best way for people to obtain all the nutrients needed for good health is to consume a balanced diet consisting of a large variety of foods.

Activate Prior Knowledge Ask students to recount the six nutrients as you write them on the board. Ask students to name foods they know are high in one of these nutrients. Under the vitamin and mineral groups, ask them to name both a food source and a specific vitamin or mineral. Review the list, prompting the students to name the food group to which each food item belongs.

2 Teach

Before Viewing the Video
For Discussion

Write all six nutrients (carbohydrates, proteins, fats, vitamins, minerals, water) on the chalkboard. Ask students to name some benefits of each nutrient; if they are not sure, provide examples: Vitamin C promotes healthy gums and teeth; calcium makes bones and teeth strong; proteins make muscles strong; carbohydrates provide energy; fats build cells and insulate the body; water cools the body. Tell them to watch closely for more clues about the six nutrients and how they contribute to good health.

After Viewing the Video
Thinking Critically

1. Have students recount the benefits of nutrients they saw in the video. Add these benefits to the appropriate chalkboard list. You may want to add extra nutrient columns (B vitamins, phosphorus) if the class responds well.
2. How can we be sure that we are getting enough nutrients? (Consume the right amount of foods from each food group by eating a diet that corresponds to the USDA Food Guide Pyramid.)

Using Video Notes

Distribute copies of Page 148 to students. Have them watch the video again. This time, have students complete the chart on their worksheet.

③ Close

Checkpoint

- What kinds of foods are in a balanced diet? (foods from all of the basic food groups)
- How do carbohydrates help your body? (provide energy)
- How do proteins help your body? (help your body grow and repair tissues)
- Why is fat an important nutrient? (provides energy; cushions bones in sockets; carries vitamins throughout body; good for nerves and blood)
- How do vitamins help your body? (D works with calcium for strong bones; A helps your bones grow and keeps your eyes healthy; E and K protect blood cells; C helps fight infections.)
- How do minerals help your body? (Calcium and phosphorus build bones and teeth; iron keeps blood healthy and helps blood carry oxygen to muscles and the brain.)
- Why is water an important nutrient? (Water contains minerals that are beneficial, carries nutrients throughout the body, carries away waste and cools the body.)

Activity

What to Expect Students will distinguish between vegetarian diets and diets with meat, and they will illustrate that it is possible to eat a healthy diet without eating meat.

Distribute copies of Page 149 and tell students to think of their favorite foods. Then ask them to choose their favorite vegetarian foods to find out whether or not they would like to be a vegetarian. If they do not end up with a balanced diet, ask them to fill in each list (using a different color pen) with the foods they would need to add to meet the food pyramid suggestions.

Note: You may want to preface this activity by discussing reasons people have for becoming a vegetarian (*cruelty to animals, health concerns, economic concerns, distaste for meat*).

FUN FACT

Did you know that you can make vitamin D just by walking into the sunlight? The sun's ultraviolet rays cause a cholesterol substance inside your body to convert into vitamin D, which is the same vitamin in milk that gives you healthy bones and teeth.

VIDEO NOTES
The Benefits of Good Nutrition

Directions Fill in this chart with (a) foods that contain each nutrient, (b) good things that we know the nutrient does for our health and (c) good things that scientists think the nutrient may do for our health.

	A Where to Find	**B** What It Does	**C** Potential Benefits
Carbohydrates			
Carbohydrates			
Fats			
Vitamins *B vitamins*			
Vitamin C			
Vitamin D			
Vitamins *Calcium*			
Vitamins *Phosphorus*			
Water			

ACTIVITY

The Benefits of Good Nutrition

Find Out What would you eat if you were a vegetarian?

What to Do

1. Using the vegetarian Food Guide Pyramid below, write down some of your favorite foods from each triangle.

2. From the list you have made, decide which foods you would want to have for breakfast, which ones for lunch, and which ones for dinner. Some foods you might even want at every meal!

3. Using your list of foods, write down your menu in the boxes according to the vegetarian Food Guide Pyramid.

Vegetarian Food Guide Pyramid

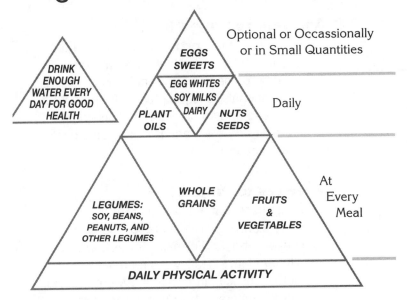

Talk About It If you were a vegetarian, would you be able to eat a balanced diet? What foods would you need to eat more of?

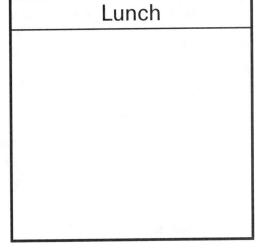

Selected Answers

Life Science

Answers to Page 8
1. organisms
2. environment
3. habitat
4. community
5. population
6. ecosystem

Answers to Page 9

Polar	African Plain
polar bear	lion
fish	wildebeest
penguin	giraffe
glacier	grassy plain

Answers to Page 12

Prey	Pred.	Scav.	Decmp.	Prod.	Cons.
worm	bird	vulture	worm	banana	monkey
insects	bird			leaves	giraffe
w'beest	cheetah			flowers	bees
				grass	w'beest

Answers to Page 16
1. structures
2. behavior
3. adaptations
4. camouflage
5. reproduction
6. inherited traits
7. learned behavior

Answers to Page 24
paper recycling: junk mail, newspaper
metal recycling: aluminum cans and foil
plastic recycling: milk container
glass recycling: drink bottles, applesauce jar
landfill or compost heap: banana peel, fish bones

Answers to Page 28
Natural disasters: forest fire, volcano eruption, earthquake, flood. More easily restored than prevented, natural disasters are a part of the life cycle that has been going on for hundreds of millions of years. Natural disasters may destroy habitats, but nature can restore itself.

Disasters caused by humans: oil spill, forest fire, chemical leak, flood. Population changes and construction usually have more lasting and more damaging effects than natural disasters. Prevention is the key because these types of disasters are so easily avoided.

Answers to Page 29
The order of the food chain follows the order of environmental restoration: small plants must be present for small animals to survive, small animals must be present for predators to survive and so on. Food and shelter must be restored before animals can come back to a damaged area.

Answers to Page 33
Responses that indicate student understanding will include variations of shelter, food, and climate. For example, a pitcher plant's healthy environment would include insects and water in a warm, humid climate, whereas its unhealthy and lethal environments would take away one or more of these requirements. *(For example, placing a pitcher plant in the desert would kill it because it would not get enough water.)*

Answers to Page 36

1. greenhouse effect, 2. Blue Whale,
3. Spotted Owl, 4. Endangered Species Act,
5. endangered, 6. rain forest, 7. global ecosystem

Earth Science

Answers to Page 44

1. closer and bigger, 2. the path a planet follows as it travels around the sun, 3. the sun, 4. at the center, 5. Mercury, Venus, Earth, Mars, 6. Jupiter, Saturn, Uranus, Neptune, Pluto

Answers to Page 45

1. Mercury, yes, yes, yes, zero
2. Venus, no, no, yes, zero
3. Earth, yes, yes, yes, 1
4. Mars, yes, yes, yes, 2
5. Jupiter, no, yes, no, 16
6. Saturn, no, no, no, 23
7. Uranus, no, no, no, 15
8. Neptune, no, no, no, 8
9. Pluto, no, no, yes, 1

Answers to Page 48

1. the Milky Way galaxy, 2. planetarium, 3. star chart, 4. stars, 5. constellations, 6. the Big Dipper, 7. solar system

Answers to Page 49

The stars actually do move, but this movement cannot be seen from Earth because they are so far away. Their apparent motion is caused by Earth's rotation.

Answers to Page 52

All answers are true except 2, 6 and 7. The correct statements are as follows:

2. Earth's upper mantle is divided into an upper zone and a lower zone.
6. The mountains that we see on Earth have formed over millions of years.
7. The crust is thicker over land and thinner over the oceans.

Answers to Page 53

Check students' work for correct relative size. *(The crust should be the thinnest layer, the mantle should be the thickest layer, and so on.)* The plates in the crust move because the lower zone of the mantle is soft and moveable while the crust is hard and rigid.

Answers to Page 56

Fast Changes: volcanoes, heavy storms, floods, earthquakes, mudslides and rockslides **Medium Changes:** rivers changing riverbanks, waves on beach **Slow Changes:** glacier movement, water, ice, deposition, winds

Answers to Page 57

1. The sand dunes were formed as the wind blew sand particles into large hills of sand.
2. The ocean arch was formed as water eroded the surrounding rock.
3. The land arch was formed as wind & water eroded away the softer surrounding rock.

Answers to Page 60

1. plateau, 2. mountain, 3. glacial plain, 4. the Grand Canyon, 5. topographic map, 6. contour lines, 7. lava plain

Answers to Page 61

Check students' work to make sure that they have charted paths in the correct elevations. Although the high elevation path may look shorter, it is probably just as long in distance *(taking into account the height of the mountains)* and would take longer because the climbs of such a hike would be more difficult.

Answers to Page 64

1. Rock cycle, 2. igneous, 3. metamorphic, 4. minerals, 5. sedimentary, 6. c, 7. b, 8. a

Answers to Page 65

Students should report that their models are like real sedimentary rocks because small pieces are cemented together. The models are different because real rocks form over long periods of time and are "glued" together by minerals and other matter.

Answers to Page 68

1. c   **2.** e   **3.** d
4. a   **5.** b
6. T   **7.** T
8. F: Soil is thinner on mountains and thicker in valleys.
9. T
10. F: Soil differs from place to place.

Answers to Page 69

1. Animals depend upon the soil for food and shelter. Animals eat plants or feed on animals that eat plants, or both.
2. These animals tunnel through the ground, so soft, moist soil is the best medium.
3. These organisms feed on nutrients and water provided by the soil.
4. The soil provides water and nutrients to underground plants and animals both directly and through the plants and animals that it feeds. A cycle is in place wherein live plants provide sustenance for other plants and animals that eventually die and decay in the soil to enrich it with nutrients for new plants to grow.

Answers to Page 72

Renewable: trees, water, air, fish, animal
Nonrenewable: minerals, soil, rocks, coal, petroleum
Inexhaustible: sunlight, wind, moving waters

Answers to Page 73

Petroleum (nonrenewable): walk, ride a bicycle, use public transportation
Trees (renewable): use the paper twice, recycle, use a computer or chalkboard
Fish (renewable): eat dolphin-safe tuna, don't eat endangered fish, don't pollute
Water (renewable): turn off water when brushing teeth, take short showers, water lawns and gardens in early evening

Physical Science

Answers to Page 76

1. a   **2.** c   **3.** b
4. b   **5.** b   **6.** a
7. c   **8.** c

Answers to Page 77

With 10 marbles, the water level should be twice the level it was with 5 marbles, and with 15 marbles, the water level should triple. As the space in the container is taken up by water and more and more marbles, the amount of air decreases at a parallel rate. Students should recognize that all three states of matter

Answers to Page 80

mixture, solution, compound, substances

Mixtures: oil and water, cereal and milk
Solutions: chocolate milk, salt water
Compounds: rust, water, tarnish, wood ashes, table salt

Answers to Page 81

oxygen, hydrogen, sodium, chlorine; water, salt; the gases combine & chemically alter to become liquid; the solid and gas combine and chemically alter to become a white solid (crystal).

Answers to Page 84

Car: gasolline, driving, mechanical
Wind-up toy: a wound-up spring, movement, mechanical
Candle: wick and wax, burning, light & heat
Match: substance on head of match and wood or paper, burning, light and heat
Campfire: wood, burning, light and heat
Flashlight: battery, flashlight shines, electrical & light and heat
Electric stove: electricity, cooking, heat
Human body: food, movement, heat & sweat

Answers to Page 85

Students should predict that filling the balloon with air will give it stored energy. When testing their predictions, students should observe the balloon moving as its air escapes, which proves that energy was used. The balloon could also be stretched or lifted to give it energy.

Answers to Page 88

1. solar energy, 2. ray, 3. opaque, 4. reflect, 5. pupil, 6. retina, 7. the sun, 8. artificial, 9. energy

Answers to Page 89

1. cornea, 2. pupil, 3. lens, 4. retina

Answers to Page 92

1. The smooth, shiny surface of a mirror reflects light straight back to us, creating a clear image.
2. Mirrors help drivers see what is behind them as well as to the side of them.
3. Answers will vary, but may include that artificial light is produced as a result of human activities, whereas natural light exists independently of human effort. Generally, natural light is synonymous with sunlight.
4. A windshield is transparent; it allows all light that strikes it to pass through. A car seat is opaque; it prevents any light from passing through.
5. It is translucent. You can't see through it, but it does allow some light to pass through.

Answers to Page 93

The reason you can see the coin after the water is poured in the cup is because the light slows down when it enters the water and is bent, or refracted, projecting the image at a different angle.

Answers to Page 96

The spectrum striking and reflected from the white snowman should contain all colors; the spectrum striking the black tire should contain all the colors but no colors should be reflected; the spectrum striking the red lips should contain all the colors but only red should be reflected.

Answers to Page 97

Students should observe that, when they spin the disk quickly, they cannot see the separate colors in the disk. Instead, the colors mix together, appearing white or grayish-white.

Answers to Page 100

1. speed, 2. motion, 3. work/force, 4. distance, 5. changing, 6. forces, 7. move, 8. machine, 9. force, 10. energy, 11. gravity

Answers to Page 101

Bike: feet pushing on pedals, pedals pulling chain
Pully: pulling to lift heavy object
Screw: push, turn past
Ax: push, wedge apart wood
Wheelchair: push, spread load over longer surface

Answers to Page 104

1. Seesaw — first class
2. Broom — third class
3. Wheelbarrow — second class
4. Canoe paddler — third class

Answers to Page 105

a. triangle
b. inclined plane
c. door hinges, furniture, kitchen appliances
d. screws and bolts differ by size, thickness, number of threads per cm, shape of the head, shape of the screwdriver slot
e. friction keeps the screw from twisting back out

Answers to Page 108

1. conductor, 2. circuit, 3. batteries, 4. heat energy, 5. light energy, 6. mechanical energy, 7. nonconductors

Answers to Page 109

5, 2, 4, 3, 6, 1

Answers to Page 113

The nail was able to pick up the paper clips because the electric current from the dry cell produced a magnetic field in the coil and the nail became a magnet. When the wire was disconnected, the circuit was broken and the electric current ceased to flow. The strength of the electromagnet could be increased by adding coils to the nail or by adding another battery.

Answers to Page 116

| 1. F | 2. T | 3. T | 4. T |
| 5. T | 6. F | 7. F | 8. F |

Answers to Page 117

The can moves and follows the balloon as long as the balloon is holding a static charge. By rubbing the balloon on your hair, you build up electrical static charge. The can rolls because the static electricity in the balloon attracts the opposite charge that is in the can. The longer you rub the balloon on your hair, the more static charge you build up. As you build up more and more charge, you build up more speed and distance.

Health Science

Answers to Page 119

Birds can fly because their hollow bones are lightweight and strong. Scientists and architects long ago figured out that the cylinder is a very strong structure for holding up weight. You may reinforce this fact by pointing out houses or buildings that use columns for support. Students will find that the strength of the cylinder is greatly weakened if the end of the tube is not even. Also, the cylinder is strongest when used in a vertical position. Relate these observations to bones and their functions *(length of bones is utilized for strength)*.

Answers to Page 120

Students should illustrate figures that:

(a) have no bones and therefore no structure

(b) have no muscles and therefore have only a skeletal shape and are unable to move

(c) have no ligaments and therefore have no connections to hold the bones together

(d) have no tendons and therefore have no connections to hold muscles to bones

Answers to Page 121

Muscles are harder when contracted because muscle fibers thicken as they shorten. They are compressed into a smaller space. Compare a piece of paper that floats in the air versus a piece of paper rolled up into a tight ball. Answers may vary, but most students should be able to flex their biceps while holding heavy books longer than they can stand on their tiptoes. However, this does not mean that the biceps are stronger than the calf muscles: the calf muscles are doing more work by holding up the greater weight of the body. Students should report that they can feel many muscles at work in each activity. Use this observation to emphasize that muscles and bones work together in groups.

Answers to Page 124

Action	Muscles and Bones
Talking	Bones: jawbone Muscles: face, tongue, lips, heart
Running	Bones: toes, feet, shins, thighs, hips, arms, neck, etc. Muscles: legs, abdomen, arms, heart
Doing Homework	Bones: hands, arms Muscles: hands, arms, face, heart
Arm Wrestling	Bones: hands, arms, shoulders Muscles: hands, arms, shoulders, abdomen, heart
Sneezing	Bones: face Muscles: neck, face, heart

Figure A.2

Answers to Page 125

Muscle	Voluntary	Involuntary
Bicep Can you flex your arm?	☒	☐
Blood Vessels Can you move your blood?	☐	☒
Eyelids Can you blink your eyes?	☒	☒
Jaw Can you yawn?	☒	☒
Toes Can you wiggle your toes?	☒	☐
Intestines Can you digest your food?	☐	☒
Thigh Can you move your leg?	☒	☐
Diaphragm Can you take a deep breath?	☒	☒
Heart Can you make your heart beat?	☐	☒
Eye Can you look all around?	☒	☐

Answers to Page 128

1. irregular
2. involuntary
3. flat
4. long
5. short
6. smooth
7. skeletal
8. cardiac
9. voluntary

Answers to Page 132

In-line skater: knee pads, helmet, wrist guards, elbow pads

Lab technician: surgical gloves, protective eye wear, apron

Lacrosse player: padded gloves, face guard, helmet

Construction worker: hard hat, protective eye wear, work gloves, back brace, ear protectors

Answers to Page 133

All answers in the chart should be "yes." The National Safety Council recommends rounded edges; cushioned ground surface *(grass, dirt and asphalt are not acceptable)*; wide openings in hanging rings, stair rungs and railings; non-rusted metal; capped nuts, bolts and clamps.

Answers to Page 136

Cut/Scrape: clean the cut, apply antibiotic ointment and bandage

Nosebleed: have the person sit down, lean slightly forward and pinch nostrils for five to ten minutes; if bleeding continues, place cold cloths on nose

Sprain: rest the sprained joint, apply ice pack to reduce swelling, compress with bandage to reduce swelling and stabilize joint in case a bone is broken, elevate the joint above the heart to reduce blood flow and pain

Answers to Page 140

2, 4, 1, 3, 5
b, c, d, c

Answers to Page 145

Each student should construct a diet that follows the USDA Food Guide Pyramid recommendations for a balanced diet. Diets should thus consist of foods from every food group, with the quantities corresponding to USDA allowances *(6-11 servings from the Bread, Cereal, Rice & Pasta Group, 2-4 servings from the Fruit Group, etc.)*. See the diagram below *(figure A.3)*.

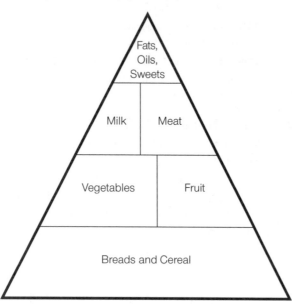

USDA Food Guide Pyramid

Answers to Page 147

Alternate Activity Check students' work and determine whether they are eating a balanced diet by making sure that they are eating the USDA recommended amount of each food group *(check the Food Guide Pyramid)*. To improve their diets, students should pinpoint specific food groups from which they are not eating enough and/or food groups from which they should eat less.

Answers to Page 149

Determine whether individual students would be able to eat a balanced diet based on their tastes for foods *(e.g., Does the student like fruits, vegetables, soy products, nuts and grains? Does the student dislike meat?)*

Notes . . .